Computer Numerical Control Programming

Computer Numerical Control Programming

Peter J. Amic

Prentice Hall

Upper Saddle River, New Jersey   Columbus, Ohio

Library of Congress Cataloging-in-Publication Data

Amic, Peter J.
Computer numerical control programming / Peter J. Amic.
p. cm.
Includes index.
ISBN 0-13-326158-1
1. Machine-tools--Numerical control--Programming. I. Title.
TJ1189.A625 1997
621.9′023--dc20 96-33099
CIP

Cover art: Courtesy of CNC Software, Inc.
Editor: Ed Francis
Production Editor: Linda Hillis Bayma
Production Coordination: Custom Editorial Productions, Inc.
Design Coordinator: Julia Zonneveld Van Hook
Cover Designer: Proof Positive/Farrowlyne Assoc., Inc.
Production Manager: Patricia A. Tonneman
Marketing Manager: Danny Hoyt

Simon & Schuster/A Viacom Company
Upper Saddle River, New Jersey 07458

Printed in the United States of America

ISBN 0-13-326158-1

Prentice-Hall International (UK) Limited, *London*
Prentice-Hall of Australia Pty. Limited, *Sydney*
Prentice-Hall Canada Inc., *Toronto*
Prentice-Hall Hispanoamericana, S. A., *Mexico*
Prentice-Hall of India Private Limited, *New Delhi*
Prentice-Hall of Japan, Inc., *Tokyo*
Simon & Schuster Asia Pte. Ltd., *Singapore*
Editora Prentice-Hall do Brasil, Ltda., *Rio de Janeiro*

Preface

Through my engineering practice and teaching experience, I have always believed that a book should be understandable to any reader with a strong desire to learn and should require only a minimum level of preparation. Therefore, CNC programming experience is not required when using *Computer Numerical Control Programming.* This book is written for a broad range of users. First, it is for beginners who wish to learn CNC programming. Those who are already operators can learn different procedures for machine setup. This book also serves as an instructionally designed textbook on CNC programming, but intermediate and advanced CNC programmers will find quick references containing conceptual ideas that work and serve as practical examples. Finally, it should help machine shop supervisors looking to improve efficiency in their CNC departments.

All concepts discussed in this text are fundamental to most CNC programming techniques. You are taken step-by-step through the creation of a program—from conception of the idea through its application, and from specifics to generalizations and back. You can be certain that every program or program segment presented in this book has been thoroughly tested and that real parts were made.

This book is organized into compact, easy-to-read, example-packed chapters. In order to accommodate the broad range of users it addresses, the book includes descriptions, applications, and illustrations, with literally hundreds of examples presented. In each chapter you will find several techniques that answer a specific application need. Key terms are in bold throughout the text and reappear in list form at the end of each chapter. Many of these terms are also defined in the book's illustrated glossary. The following is a description of each chapter and appendix.

- Chapter 1, "Fundamentals of Computer Numerical Control," explains how CNC machines work.
- Chapter 2, "Programming Fundamentals," touches on all topics related to successful creation of CNC programs.
- Chapter 3, "Machine Setup," is a conceptual and practical look at one of the most important and often troublesome aspects of CNC features. It is especially designed to answer the questions that CNC operators encounter.
- Chapter 4, "Manual Tool Radius Compensation," discusses the need for tool radius compensation when programming. This chapter also teaches how knowledge in mathematics solves real programming problems.
- Chapter 5, "Automatic Tool Radius Compensation," is an overview of powerful built-in machine software. Comprehensive examples and tips serve as preparation for the advanced stages of CNC programming.
- Chapter 6, "Tooling Features," discusses the tooling on CNC machines, as well as cutting conditions.
- Chapter 7, "Programming CNC Lathes," and Chapter 8, "Programming CNC Machining Centers," present advanced topics for programming the lathe or machining center. Both novices and experienced programmers will benefit from these chapters.
- Chapter 9, "Macro Programming," is a conceptual and practical look at the most advanced stage of CNC programming. It gives programmers step-by-step information for creating successful macros and becoming more productive.
- Chapter 10, "Using Personal Computers in CNC Programming," describes the role of personal computers in creating and storing programs. It also illustrates the powerful features of part-programming software and discusses the features of APT and COMPACT II computer programming languages.

Computer Numerical Control Programming also contains appendices offering additional tools for learning about CNC programming. Appendix A presents a list of CNC codes, distinguishing between the modal and nonmodal codes, as well as the address description. Appendix B provides the most-used formulas when programming different machining operations, using either inch or metric data. Appendix C presents the compensation amount when chamfering on the lathe, calculated for the most-used angles and tool nose radius values. Appendix D presents general notes for safe CNC operations. Appendix E presents the answers to chapter self-tests.

The situations discussed throughout the book represent the type of problems encountered in actual practice. Your mission, should you decide to accept it, is to complete all of the self-tests in the book. Once you do this successfully, you will have acquired basic knowledge of CNC programming, and you can go on to master the concepts by experimenting on your own. If you are an instructor, you may wish to include these exercises, as well as Relating the Concepts questions, in student assignments.

Acknowledgments

I would like to acknowledge and thank the following individuals and firms for their contributions and assistance on this project: James J. Bushong, Hitachi Seiki U.S.A., Inc., Congers, New York; Thomas Carlberg, Sandvik Coromant Company, Fair Lawn, New Jersey; Brent Daley, University in Calgary, Alberta, Canada; Michelle Gruda, Valenite Inc., Madison Heights, Michigan; Jim Harkins, Northern Alberta Institute of Technology, Edmonton, Alberta, Canada; Robert T. Kogan, Sandvik Coromant Company, Fair Lawn, New Jersey; Ben Mund, CNC Software, Inc., Tolland, Connecticut; Donna Thompson, Mitsubishi Materials U.S.A. Corporation, California; and D. Tichenor, DT Community College.

I also wish to thank the reviewers of the manuscript: James B. Higley, Purdue University Calumet; Richard A. Kruppa, Bowling Green State University, Bowling Green, Kentucky; Andrew Paris, Highling Community College; Terry Thomas, Jackson Community College; and Bruce C. Whipple, Trident Technical College.

Contents

1 Fundamentals of Computer Numerical Control

Key Concepts

NC and CNC

The Applications of NC/CNC

The Construction of CNC Machines
- Computers
- Drive Motors
- Control Systems
- Tool Changers

The Coordinate Systems
- The Rectangular Coordinate System
- The Polar Coordinate System

The Grid System

Reference Points
- Machine Origin
- Part Origin
- Program Origin

Coding Systems

CNC Syntax
- The Computer Word Address Format
- The End of Block Code

Types of CNC Codes

NC and CNC

Numerical Control (NC) is any machining process in which the operations are executed automatically in sequences as specified by the program that contains the information for the tool movements. The NC concept was proposed in the late 1940s by John Parsons of Traverse City, Michigan. Parsons recommended a method of automatic machine control that would guide a milling cutter to produce a "thru-axis curve" in order to generate smooth profiles on work pieces. In 1949, the U.S. Air Force awarded Parsons a contract to develop a new type of machine tool that would be able to speed up production methods. Parsons commissioned the Massachusetts Institute of Technology (M.I.T.) to develop a practical implementation of his concept. Scientists and engineers at M.I.T. built a control system for a two-axis milling machine that used a perforated paper tape as the input media. In a short period of time, all major machine tool manufacturers were producing some machines with NC, but it was not until the late 1970s that computer-based NC became widely used. NC matured as an automation technology when inexpensive and powerful microprocessors replaced hard-wired logic-making computer-based NC systems.

When Numerical Control is performed under computer supervision, it is called **Computer Numerical Control (CNC).** Computers are the control units of CNC machines. They are built in or linked to the machines via communications channels. When a programmer programs, say, taper or radius, he or she enters some information in the program, but the computer calculates all necessary data to get the job done. This book uses the term *CNC,* because these days almost every NC system is controlled by computer.

On the first Numerically Controlled (NC) machines, numerical data was controlled by tape, and because of that, the NC systems were known as tape-controlled machines. They were able to control a single operation entered into the machine by punched or magnetic tape. There was no possibility of editing the program on the machine. To change the program, a new tape had to be made.

Today's systems have computers to control data; they are called *Computer Numerically Controlled (CNC)* machines. For both NC and CNC systems, work principles are the same. Only the way in which the execution is controlled is different. Normally, new systems are faster, more powerful, and more versatile.

The Applications of NC/CNC

Since its introduction, NC technology has found many applications, including lathes and turning centers, milling machines and machining centers, punches, electrical discharge machines (EDM), flame cutters, grinders, and testing and inspection equipment. The most complex CNC machine tools are the turning center, shown in Figure 1–1, and the machining center, shown in Figure 1–2.

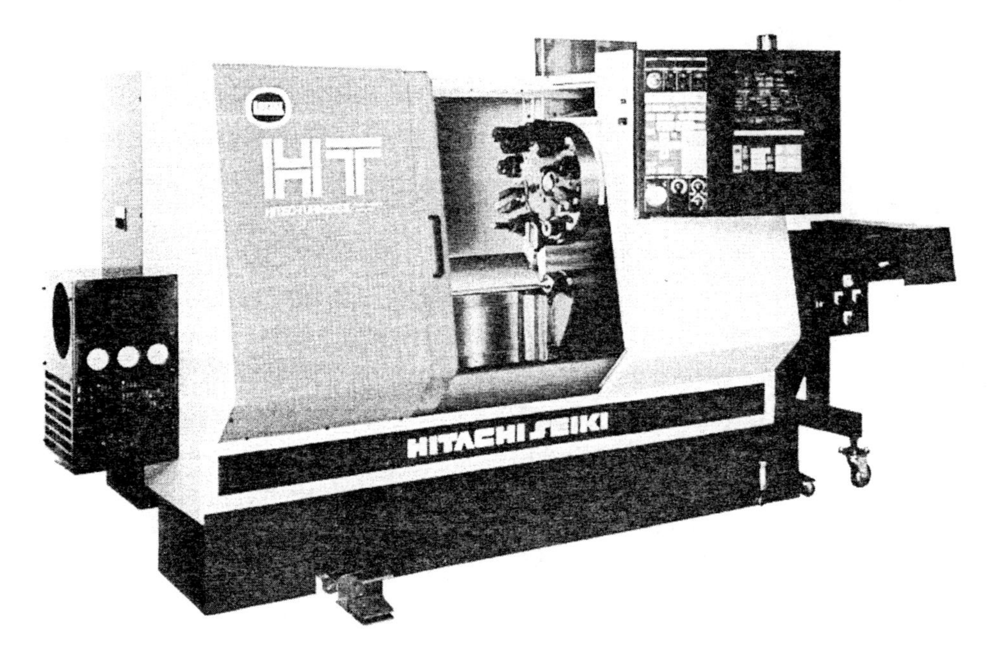

Figure 1–1 A modern turning center with a ten-station turret that accepts quick-change tools. Each tool can be positioned in seconds with the press of a button. (Courtesy Hitachi Seiki U.S.A., Inc.)

Figure 1–2 A heavy-duty horizontal machining center with "cell-ability," which allows pallet pools, a rail-guided pallet system, and additional tool change capability to be added at any time. (Courtesy Hitachi Seiki U.S.A., Inc.)

When preparing a program for a particular operation, the programmer must select all cutting data using recommendations for conventional machining. This includes proper selection of cutting speeds, feed rates, tools and tool geometry, and so on. When the programmer has chosen all of the necessary information properly, the operator loads the program into the machine and presses a button to start the cutting cycle. The CNC machine moves automatically from one machining operation to another, changing the cutting tools and applying the coolant. In a surprisingly short time, the workpiece is machined according to the highest quality standards (Figure 1–3). But that is not all! No matter how big the work series is, all of the parts will be almost identical in size and surface finishing. At this time of advanced technology, with its high demands for surface finishing and tolerances of components in, for example, aerospace, nuclear, and medical equipment manufacturing, only CNC machines provide successful results.

CNC machines have many advantages over conventional machines. Some of them are:

1. There is a possibility of performing multiple operations on the same machine in one setup.
2. Because of the possibility of simultaneous multi-axis tool movement, special profile tools are not necessary to cut unusual part shapes.
3. The scrap rate is significantly reduced because of the precision of the CNC machine and lesser operator impact.

Figure 1–3 Typical oilfield parts made on CNC machines. (Courtesy Stratex Drill Systems.)

4. It is easy to incorporate part design changes when CAD/CAM systems are used.
5. It is easier to perform quality assurance by a spot-check instead of checking all parts.
6. Production is significantly increased.

CNC machines also have some disadvantages:

1. They are quite expensive.
2. They have to be programmed, set up, operated, and maintained by highly skilled personnel.

Obviously, CNC machines have more advantages than disadvantages. The companies that adopt CNC technology increase their competitive edge. As always with new technology, the cost per CNC unit is being cut further and more companies can afford CNC equipment, which enables them to answer the increasingly strong requirements for production speed and quality that competitive markets demand. In the future, the broader use of CNC machines will be one of the best ways to enhance automation in manufacturing.

The Construction of CNC Machines

CNC machine tools are complex assemblies, and a more detailed study is a topic for a separate book. However, in general, any CNC machine tool consists of the following units:

1. Computers
2. Control systems
3. Drive motors
4. Tool changers

According to the construction of CNC machine tools, CNC machines work in the following (simplified) manner:

1. The CNC machine control computer reads a prepared program and translates it into machine language, which is a programming language of binary notation used on computers, not on CNC machines.
2. When the operator starts the execution cycle, the computer translates binary codes into electronic pulses which are automatically sent to the machine's power units. The control units compare the number of pulses sent and received.
3. When the motors receive each pulse, they automatically transform the pulses into rotations that drive the spindle and lead screw, causing the spindle rotation and slide or table movement. The part on the milling machine table or the tool in the lathe turret is driven to the position specified by the program.

Computers

The first NC machines relied on electronic hardware based on the digital circuit technology available at that time. These machines had no memory and were not able to store programs. To produce a new part, the NC machine had to reread the program one statement at a time and execute each statement before proceeding. CNC machines, introduced in the late 1970s, are less dependent on hardware and more dependent on software. These machines store a program into memory when it is first read in. This provides for faster operation when producing a number of identical parts, since the program can be recalled from memory repeatedly without having to read it again. CNC machines use an on-board computer that allows the operator to read, analyze, and edit programmed information, while NC machines require operators to make a new tape to alter a program. In essence, the computer is what distinguishes CNC from NC.

As with all computers, the CNC machine computer works on a binary principle using only two characters, 1 and 0, for information processing. The computer reacts on precise time impulses from the circuit. There are two states, a state with voltage, 1, and a state without voltage, 0. Series of ones and zeroes are the only states that the computer distinguishes. Called *machine language,* it is the only language the computer understands. When creating the program, the programmer does not care about the machine language; he or she simply uses a list of codes and keys in the meaningful information. Special built-in software compiles the program into machine language and the machine moves the tool by its servomotors. However, the programmability of the machine is dependent on whether there is a computer in the machine's control. If there is a minicomputer programming, say, a radius (which is a rather simple task), the computer will calculate all the points on the tool path. On the machine without a minicomputer, this may prove to be a tedious task, since the programmer must calculate all the points of intersection on the tool path.

Modern CNC machines use 32-bit processors in their computers that allow fast and accurate processing of information. This results in a savings of machining time.

Drive Motors

The drive motors control the machine slide movement on NC/CNC equipment. They come in four basic types:

1. Stepper motors
2. DC servomotors
3. AC servomotors
4. Fluid servomotors

Stepper motors convert a digital pulse, generated by the microcomputer unit (MCU), into a small step rotation. Stepper motors have a certain number of steps that they can travel. The number of pulses that the MCU sends to the stepper motor controls the amount of the rotation of the motor. Stepper motors are mostly used in applications where low torque is required.

Stepper motors are used in open-loop control systems, while AC, DC, or hydraulic servomotors are used in closed-loop control systems. (Control systems are discussed in the next section.)

Direct current (DC) servomotors are variable speed motors that rotate in response to the applied voltage. They are used to drive a lead screw and gear mechanism. DC servos provide higher-torque output than stepper motors.

Alternative current (AC) servomotors are controlled by varying the voltage frequency to control speed. They can develop more power than a DC servo. They are also used to drive a lead screw and gear mechanism.

Fluid, or hydraulic, servomotors are also variable speed motors. They are able to produce more power, or more speed in the case of pneumatic motors, than electric servomotors. The hydraulic pump provides energy to valves that are controlled by the MCU.

Control Systems

There are two types of control systems on NC/CNC machines: open loop and closed loop. The overall accuracy of the machine is determined by the type of control loop used.

The **open-loop control system** does not provide positioning feedback to the control unit. The movement pulses are sent out by the control and they are received by a special type of servomotor called a stepper motor. The number of pulses that the control sends to the stepper motor controls the amount of the rotation of the motor. The stepper motor then proceeds with the next movement command. Since this control system only counts pulses and cannot identify discrepancies in positioning, the control has no way of knowing that the tool did not reach the proper location. The machine will continue this inaccuracy until somebody finds the error.

The open-loop control can be used in applications in which there is no change in load conditions, such as the NC drilling machine. The advantage of the open-loop control system is that it is less expensive, since it does not require the additional hardware and electronics needed for positioning feedback. The disadvantage is the difficulty of detecting a positioning error.

In the **closed-loop control system,** the electronic movement pulses are sent from the control to the servomotor, enabling the motor to rotate with each pulse. The movements are detected and counted by a feedback device called a *transducer.* With each step of movement, a transducer sends a signal back to the control, which compares the current position of the driven axis with the programmed position. When the number of pulses sent and received match, the control starts sending out pulses for the next movement.

Closed-loop systems are very accurate. Most have an automatic compensation for error, since the feedback device indicates the error and the control makes the necessary adjustments to bring the slide back to the position. They use AC, DC, or hydraulic servomotors.

Tool Changers

Most of the time, several different cutting tools are used to produce a part. The tools must be replaced quickly for the next machining operation. For this reason, the majority of NC/CNC machine tools are equipped with **automatic tool changers,** such as magazines on machining centers and turrets on turning centers (Figure 1–4). They allow tool changing without the intervention of the operator. Typically, an automatic tool changer grips the tool in the spindle, pulls it out, and replaces it with another tool. On most machines with automatic tool changers, the turret or magazine can rotate in either direction, forward or reverse.

Tool changers may be equipped for either random or sequential selection. In **random tool selection,** there is no specific pattern of tool selection. On the machining center, when the program calls for the tool, it is automatically indexed into waiting position, where it can be retrieved by the tool handling device. On the turning center, the turret automatically rotates, bringing the tool into position.

In random tool selection, the tools do not have to be loaded into the magazine or turret in the order in which they are called for in the program. The machine control knows where to find a particular tool as many times as it appears in the program. For instance, in one program, tools are called for in the following order: 1-2-3-4-5. However, they may be loaded in the magazine in any order, such as 5-1-3-4-2. This is illustrated in Figure 1–5.

In **sequential tool selection,** the tools must be loaded in the exact order in which they are called for in the program (Figure 1–6). Even if the tools are not in the correct order, the next tool is automatically selected, whether or not it is suitable for the next machining operation. When it is necessary to use a tool twice, the operator must load another tool with the same purpose.

The advantage of sequential tool selection is that less time is needed for indexing the tool into waiting position. The disadvantage is that more time is needed

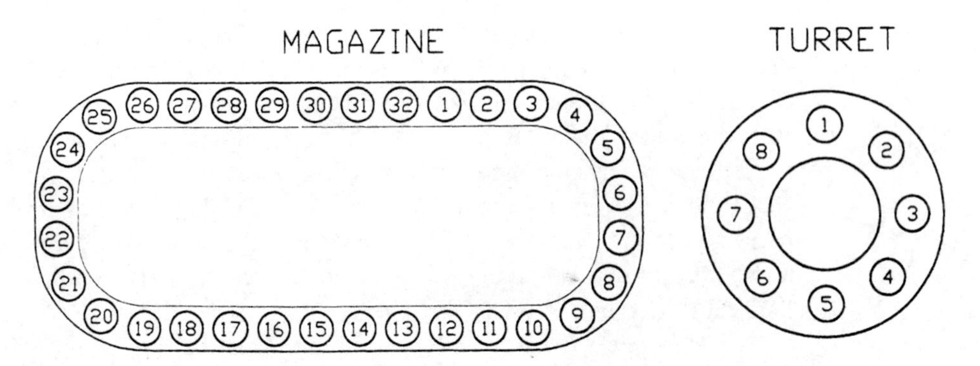

Figure 1–4 The automatic tool changers used most often.

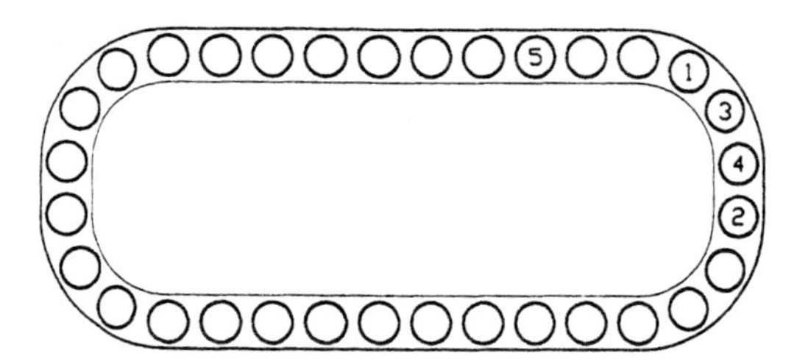

Figure 1–5 Random tool selection.

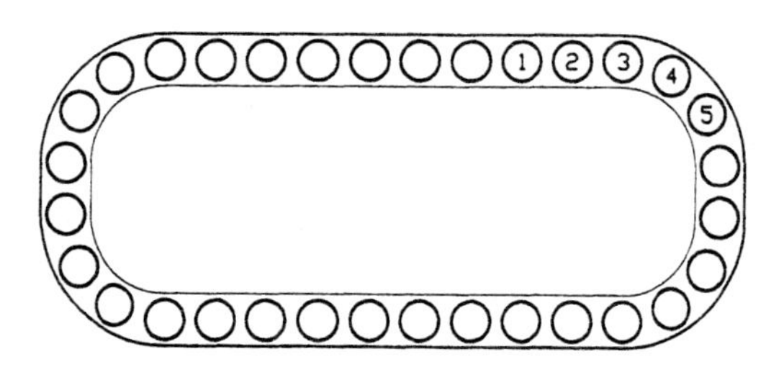

Figure 1–6 Sequential tool selection.

for setup when switching to a job with a different order of tools. This means that although the same tools are to be used, they have to be preloaded (rearranged) because of a different order in the program. The majority of modern machines are able to return the tool in the magazine and to search for the next tool during the program execution. This eliminates the time advantage of sequential tool selection, making random tool selection a standard feature on today's CNC machine tools.

The Coordinate Systems

CNC machine tools use two basic types of programming when producing a part: positioning or point-to-point, and contouring or continuous-path.

Positioning or **point-to-point programming** is used to move the tool from one point to another when programming operations such as drilling, tapping, boring, and reaming (Figure 1–7). After the operation is finished at one location,

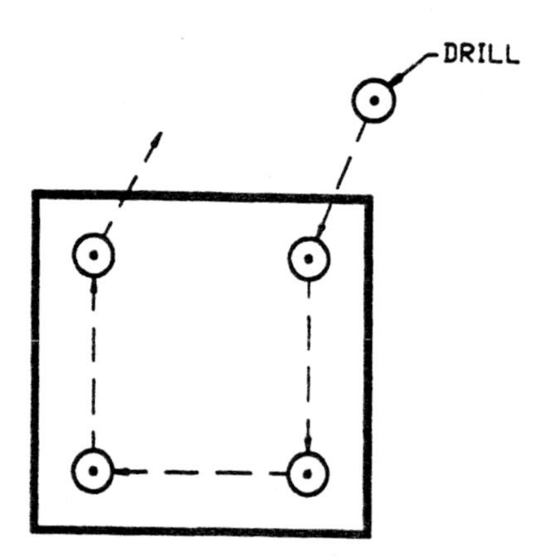

Figure 1–7 Point-to-point movements.

the tool travels to another. The main characteristic of positioning is that the tool is not in constant contact with the part. The rate of movement is usually the machine's maximum.

In **contouring** or **continuous path programming,** the tool is in constant contact with the part until the operation is finished (Figure 1–8). The tool is moved by the specified feed rate, not by rapid traverse, as in positioning. Milling and turning are the most common operations when contouring.

Either using contouring or positioning, the programmer must program the tool to move in a known direction. The direction depends on where the tool is moving: up or down, left or right. The location at which the tool finishes its travel is known as a *coordinate point.* This point is a distance away from the starting point. Thus, a coordinate point may be described by the distance and direction in which the tool travels.

In mathematical terms, any specific point can be described by distance and direction. This concept is used in plane geometry through the applications of the coordinate system, and has been accepted for CNC machine tools. There are two coordinate systems used on CNC machines: the rectangular, also known as Cartesian, and the polar coordinate systems.

The Rectangular Coordinate System

The **rectangular coordinate system** is a standard system used on CNC machines, and all of the machine movements are based on this system. One dimension is length, and the other is width. It is also the way the parts are shown on the blueprints, even though they are three-dimensional objects.

The rectangular coordinate system has two perpendicular axes on a two-dimensional surface, which is known as a *plane*. One axis is generally recognized as the horizontal axis. The line perpendicular to it is known as the vertical axis.

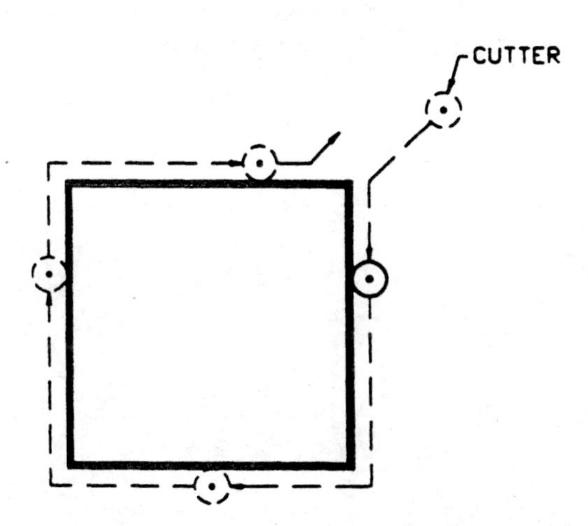

Figure 1–8 Continuous path movements.

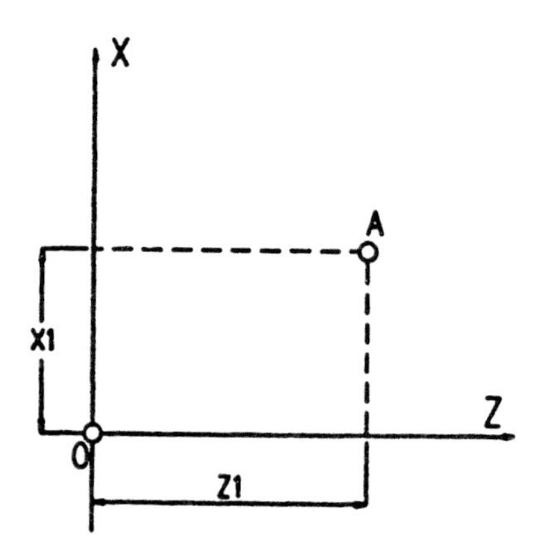

Figure 1–9 The coordinate system on the CNC lathe.

The X and Z axes are used to define the tool positions on the CNC lathe coordinate system, as shown in Figure 1–9. On the lathe, the horizontal axis of the machine is the Z axis. It is the spindle centerline. The X axis, a line perpendicular to the Z axis, is recognized as the vertical axis. The point where the axes intersect the coordinate system zero point is called a *datum* or an *origin.* All points are defined according to that position. Thus, a position of point A in Figure 1–9 is defined by the distances X1 and Z1 from the origin. (We could use any suitable numbers instead of the X1 and Z1.)

The same coordinate system is used to define the positions of the tool path on the CNC machining center, but the axes are named differently. They are the X and Y axes. An additional axis known as the Z axis is necessary to express the depth (Figure 1–10).

On the machining center, the X axis parallels the length of the table and the Y axis is perpendicular to it. The Z axis represents the spindle centerline. This axis

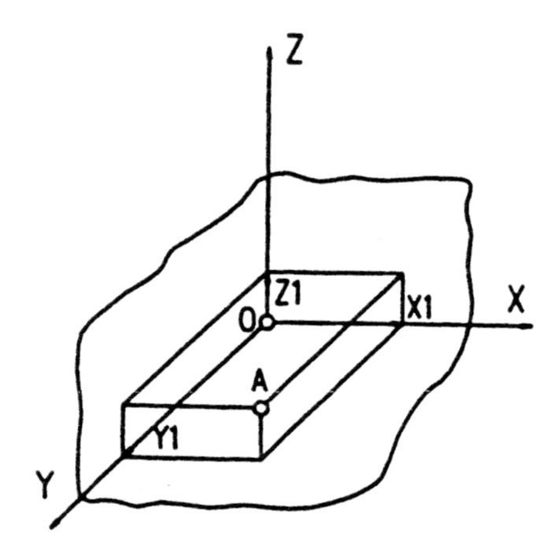

Figure 1–10 The coordinate system on the CNC machining center.

gives a necessary third dimension to this machine. Thus, point A in Figure 1–10 is defined by the distances X1, Y1, and Z1 from the origin.

The rectangular coordinate system is divided into the four sections called *quadrants*. They are numbered from I to IV in a counterclockwise direction. Quadrants of the lathe coordinate system are shown in Figure 1–11.

As shown, the X coordinates have positive values from the origin-up and negative values from the origin-down. The Z coordinates have positive values from the origin-right and negative values from the origin-left. As an exercise, we will write the coordinates for the points shown in Figure 1–12.

According to the coordinate system zero point, these points, which are in different quadrants, are defined as follows:

PT1 X3.0 Z4.0 PT2 X4.0 Z-2.0 PT3 X-4.0 Z-3.0 PT4 X-2.0 Z3.0

On the machining center, the X coordinates are positive from the origin-right and negative from the origin-left. The Y coordinates are positive from the origin-up and negative from the origin-down. The Z coordinates are positive from the origin-up and negative from the origin-down. The quadrants of the machining center coordinate system are shown in Figure 1–13. Thus, if we move the tool 5 inches down on the Z axis, we would enter Z-5.0. If the move is in an upward direction, it would be programmed as a positive value, Z5.0.

This coordinate system follows a rule of the right-hand coordinate system. Some CNC lathes use the opposite coordinate system, known as the left-hand coordinate system. The main characteristic of the left-hand coordinate system is that the sign for the X axis is minus from the origin up and plus from the origin-down. It only affects a sign for some addresses, but in reality it does not make any significant difference when programming and operating the machine. (Just make sure you know what type of coordinate system you are using.)

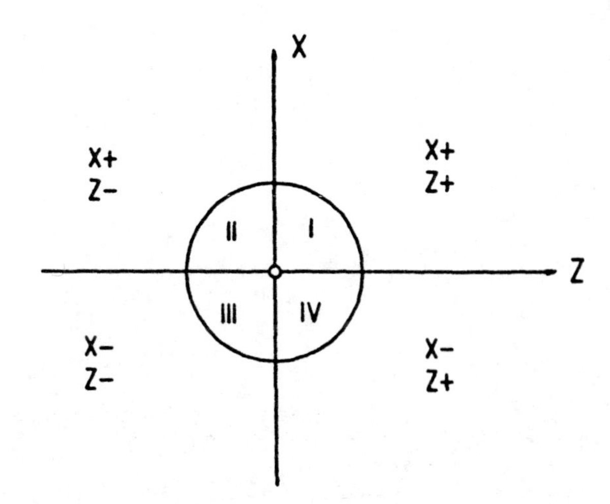

Figure 1–11 Quadrants of the lathe coordinate system.

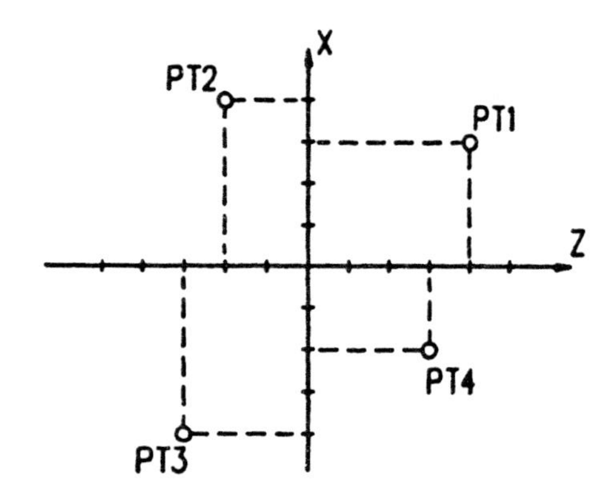

Figure 1–12 Defining points in the lathe coordinate system.

The Polar Coordinate System

On some machines the tool position can be defined using the **polar coordinate system.** It is used to easily program hole positions on a bolt circle. On the polar coordinate system, a point is defined by the radius vector and angle. The radius vector is the distance from the point to the origin. The angle is measured from the positive side of the X axis (Figure 1–14).

Usually, a machine that has the polar coordinate system also has the rectangular coordinate system. This enables programming in either coordinates, rectangular or polar. Some controls are able to accept direct polar coordinate input and automatically calculate the rectangular coordinates. When the machine does not have such capabilities, the programmer converts the polar coordinates into rectangular coordinates using trigonometric functions. For instance, to find the rectangular coordinates of point A in Figure 1–14, use the following expressions:

$$X = R \cdot \cos \alpha$$
$$Y = R \cdot \sin \alpha$$

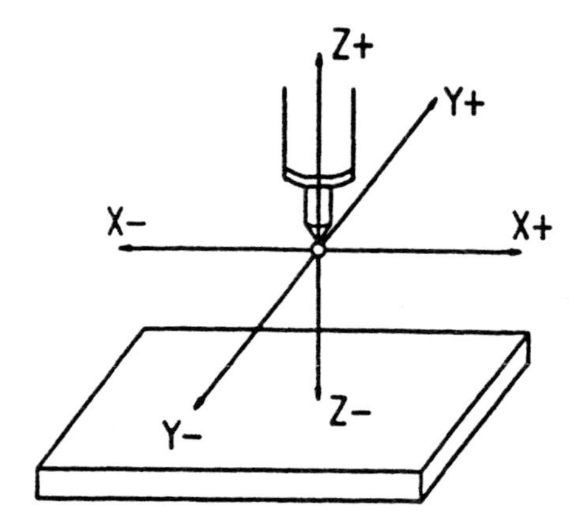

Figure 1–13 Quadrants of the machining center coordinate system.

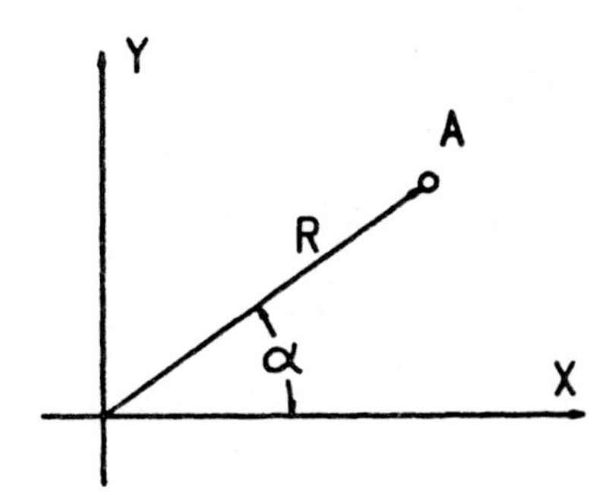

Figure 1–14 The polar coordinate system.

To find the rectangular coordinates of any point defined by the radius and angle, the same expressions apply. For instance, with a 2.0-inch radius and a 30-degree angle, we make the following calculations using the values of the trigonometric functions for a 30-degree angle:

$$X = 2.0 \bullet \cos 30 = 2.0 \bullet 0.866 = 1.732$$
$$Y = 2.0 \bullet \sin 30 = 2.0 \bullet 0.5 = 1.0$$

When comparing these values, notice that the tool travels more on the X axis than on the Y axis. This is because the angle is less than 45 degrees. It would be opposite if the angle were greater than 45 degrees. Using this simple way of checking calculations, the programmer can detect a rough error that sometimes happens when he or she accidently interchanges the sine and cosine in equations.

When using the polar coordinate system, the programmer should always enter the angle defined from the positive part of the X axis. For instance, if the radius vector splits the second quadrant, it makes a 45-degree angle with the negative X axis, but the programmer should use a 135-degree angle (90 + 45) since it is related to the positive X axis. The amount of the tool motion would be the same whether we enter a 45- or 135-degree angle, but the direction in which the tool travels would not be the same. In the first case the tool moves through the first quadrant; in the second case it moves through the second quadrant. This means that when using the polar coordinate system, the programmer can describe a point through the distance and direction, in the same way as when using the rectangular coordinate system.

The Grid System

When the tool is moved on the CNC machine, either manually or by programmed instruction, how is its travel measured? All the axes have units of measure, also known as the *grid.* The size of the grid depends on the machine resolution or increment. A control system with the resolution of 0.001 inch forms an electronic grid incremented every 0.001 inch (Figure 1–15).

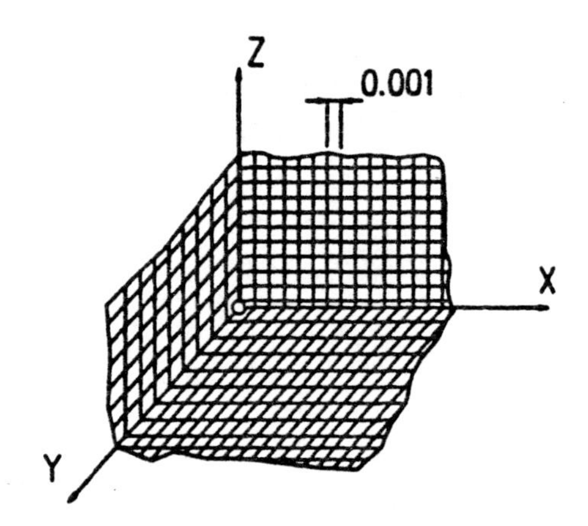

Figure 1–15 Electronic grid.

The tool always moves in the smallest increment, or the smallest amount of movement that the CNC machine can perform. The increment depends on the resolution of the machine. If the increment of one particular machine is 0.0001 inch, then the servomotor, which drives the tool, has to receive 10,000 impulses to move the tool 1 inch. In reality, the tool moves 10,000 times in increments of 0.0001 inch, although it looks like the tool moves in one smooth motion.

Machines with higher resolution have smaller distances between the increments. A resolution of 0.0001 inch is today's standard. Some special machines, such as CNC grinders, have an even higher resolution.

Most machines can operate in either the inch or metric mode. When using metric mode, the smallest input increment is 0.001 mm. It is smaller than 0.0001 inch, because 0.001 / 25.4 = 0.000039 inch. The machine can then be targeted to the destination points with a finer location. However, there is no difference in preciseness when using either inch or metric mode.

How do you know if the machine is set to use inch or metric units? The simplest way is to look at the machine monitor. If the coordinates are expressed with three decimal places, the machine is set to use metric input. If there are four digits after the decimal point, the machine is set to use inch input, as illustrated in the following examples:

Z3.1250 (The machine is set to use inch data input.)
X10.450 (The machine is set to use metric data input.)

Note that these are examples for the newer machines. On the older machines one less decimal place may be used.

Reference Points

When programming and setting up the machine, it is necessary to establish some meaningful reference points. (In principle, the same reference points are used on all CNC machine tools.) One of the most important advantages of the CNC machine tool is that it is up to the programmer and the operator where to set the part coordinate system and its zero point. Then the reference points can be established as needed.

Machine Origin

The **machine origin** or **machine zero point** is a fixed point set by the machine tool builder. Usually, it cannot be changed. Any tool movement is measured from this point. The control always remembers the tool distance from the machine origin. On the machine monitor, this distance is shown as the machine distance or relative distance. When setting up the machine, this information is used to establish the work coordinates and the tool change position.

On the lathe, the machine origin is usually set at the top corner on the right-hand side of the machine. It is the end point of the tool travel for both the X and the Z axes in the positive direction. On the machining center, the machine origin for the X and Y axes is most often set at the top corner on the right-hand side of the table. For the Z axis, it is usually set at the tool change position.

Returning the machine to its origin point is known as *zeroing the slides.* The operator always zeroes the slide in the plus direction on each axis. He or she can achieve this by setting the mode selector on the operator panel to the Zero Return position. If the mode selector is not set this way, the overtravel alarm will occur. The operator can eliminate the alarm by moving the machine in the opposite direction and by pressing the RESET button when the mode selector is on Edit position. Note that the term "moving the machine" means moving the machine slide with the tool manually.

After returning the machine to its origin, the operator enters the zero values for the X and Z axes on the lathe, or X, Y, and Z axes on the mach˙ning center. This helps to accurately establish other reference points. On some machines, the operator can't start work until he or she zeroes the slides after switching the machine on—a good safety feature.

Part Origin

The **part origin** or the **part zero point** can be set at any point inside the machine's electronic grid system limits. Establishing the part origin is also known as *zero shift, work shift, floating zero,* or *datum.* It is the origin of the absolute coordinate system.

Usually, the part origin must be established for each new setup. Zero shifting allows the relocation of the part origin to another location. It enables the operator to set the part where it is desired on the machining center table, as well as to machine long or short parts on a lathe.

All newer machines are designed to perform full zero shifting. Older machines are fixed zero machines, which have a permanent zero location that does not allow

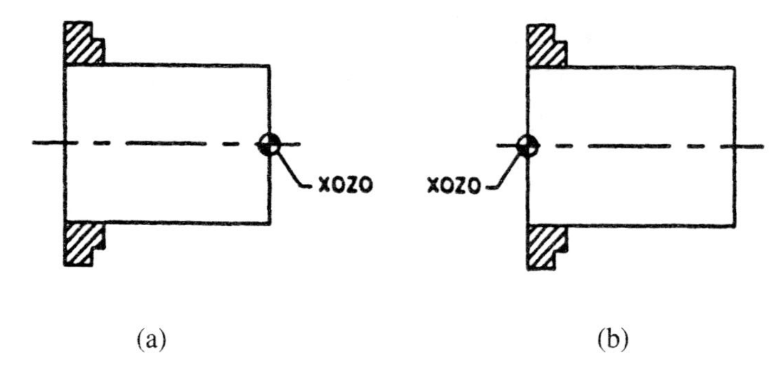

Figure 1–16 The part origin for the lathe work must be at the spindle centerline.

any change in a part zero position. On these machines, zero shifting is still possible by using a special feature called *offset.*

On the lathe, the part origin is usually set at the part finished front face. When the part is dimensioned from the back face, it is more convenient to use that side to set the part origin. Both cases are shown in Figure 1–16. In either way, a part origin for the lathe work must be on the spindle centerline.

On the machining center, the part origin can be also established at any convenient point on the part or even out of the part in the physical and electronic limits of the machine. Some examples are shown in Figure 1–17.

At position (a), the part origin is set at the centerline of the top face. This is the most frequently used way to establish the part origin on this machine. Many jobs

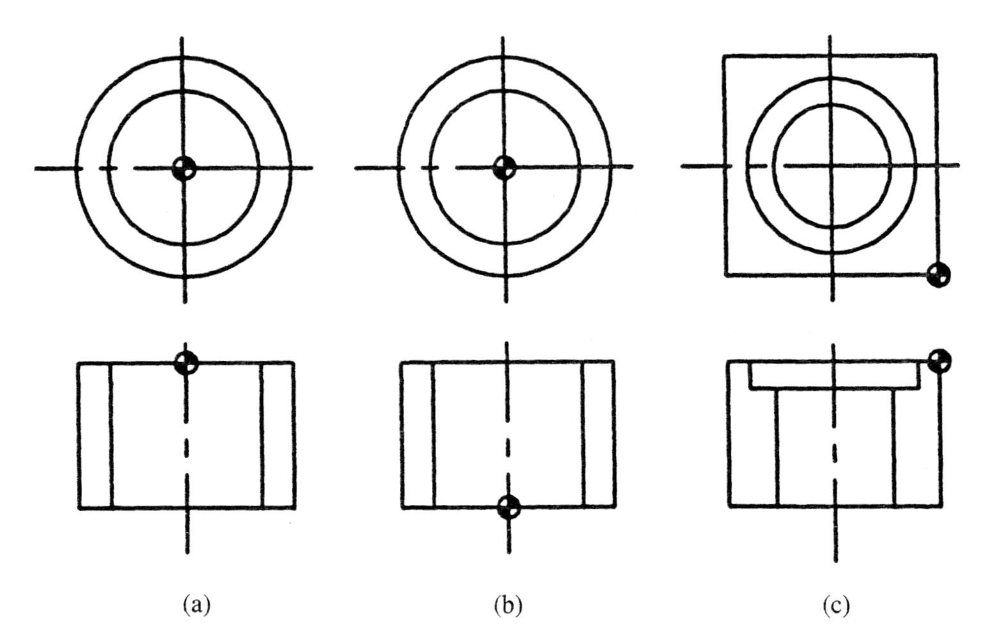

Figure 1–17 Part origin on the machining center.

done on the machining center are done in relation to the hole center, such as drilling, countersinking, and tapping on the top surface. At position (b), the part origin is set at the centerline of the bottom face. Some back boring operations are more easily programmed in this way. At position (c), the part origin is set at the intersection of two outside surfaces. Contour milling operations may be programmed using this way of setting the part origin. These examples show how flexible the CNC system is.

Part accuracy may depend on where the part origin is set. It should be set in such a way that the important sizes can be machined within the tolerances. For instance, if holes on a bolt circle are to be machined, concentricity with the bore may be required. Then it is normal to set the part origin at the bore center. When deciding where to set the part origin, study the blueprint carefully. The notes such as *parallel with, concentric within,* and so on may help the programmer choose the best location.

Program Origin

In order to start machining, it is not enough to design the program, load the tools, and establish the part origin. Consider lathe work: some parts are longer, some are shorter, and tools differ in length. At the tool start point, the control must be told where the tool is according to the part origin. It enables the control to keep track of the tool at all times, even when cutting in automatic mode is interrupted by manual operations.

The tool start point is the **program origin.** It is related to the part origin since the tool is a certain distance away from the part origin when the tool change occurs. Thus, the program origin may be described as the distance from the tool tip to the part origin when the tool is at the tool change position.

Within the program, the program origin is usually assigned using the coordinate system preset codes: G50 for the lathe and G92 for the machining center. On some lathes, code G92 is used instead of code G50 to preset the coordinate system. In general, the coordinate system preset is programmed as the first block for each tool. If not, it must be programmed before any move command. The following are examples of establishing the program origin:

Example 1 Establishing the lathe coordinate system.

N1 G50 X15.0 Z3.0 (This means that the distance from the lathe tool tip to the part origin is 15.0 inches on the X axis, and 3.0 inches on the Z axis.)

Example 2 Establishing the machining center coordinate system.

N1 G92 X20.0 Y10.0 Z8.0 (This usually means that the distance from the center of the machining center tool to the part origin is 20.0 inches on the X axis, 10.0 inches on the Y axis, and 8.0 inches on the Z axis. The distance is measured from the tool change position.)

The values entered in the coordinate system preset block describe the program origin for one particular tool. In order to return the tool to its start position, the programmer must enter the same values programmed in the coordinate system preset block in the return command.

Coding Systems

In order for the control to interpret and execute programmed information, the programmer and the operator must use a coding system. A frequently used coding system is the Binary-Coded Decimal, or BCD, system. This system is also known as the **EIA code set** because it was developed by the Electronics Industries Association. The newer coding system is the American Standard Code for Information Interchange, or ASCII. Because of its wide acceptance, it has became the International Standards Organization **(ISO) code set.**

Older NC machines without the control computer and memory use the EIA code set only. Today's CNC machines are able to use both code sets. Each of these code sets consists of the following characters used in communication with the CNC machine:

1. Numerals from 0 to 9
2. Letters from A to Z (upper and lower case)
3. Operators: Plus (+), Minus (–), Equal (=), Multiply (•), Divide or slash (/)
4. Special characters:

   Brackets () Not assigned in EIA code set
   Decimal point (.)
   Comma (,)
   Colon (:) Not assigned in EIA code set
   Percentage (%)
   Dollar sign ($)
   Hash sign (#)
   Space
   Carriage return (CR)
   Line feed (LF)
   TAB

A format error or alarm will occur if the programmer or the operator enters any code out of the code set when communicating with the CNC equipment. On some older machines, it is not possible to load a complete program from the office computer if the programmer enters any character out of the code set. Program reading stops just before the wrong character.

The most common errors in communicating with the CNC machine are typing errors. The following lists some of them:

XZ (any two addresses entered together)
Y1 5 (space between the characters)

- -1.0 (minus sign entered twice)
\ (backslash instead of slash)
O (letter O instead of number 0)

To continue program loading or program execution, the error has to be fixed, but it is usually hard to find this kind of error.

CNC Syntax

The CNC machines use a set of rules to enter, edit, receive, and output data. These rules are known as *CNC syntax, programming format,* or *tape format.* The format specifies the order and arrangement of information entered. This is an area where controls differ widely. There are rules for the maximum and minimum numerical values and word lengths that can be entered, and the arrangement of the characters and computer words is important. If the control finds any discrepancy according to these rules, it will display the message "Format error," or older machines will give an alarm message. There is a simple reason for this: It does not know what the programmer or the operator wants it to do. To continue, he or she must fix the error. This is a good safety feature.

The Computer Word Address Format

The most common CNC format is the **word address format.** (The other two formats are the fixed sequential block address format and tab sequential format. Both are obsolete.) The instruction block consists of one or more words. A word consists of an address followed by numerals. For the address, one of the letters from A to Z is used. The address defines the meaning of the number that follows the address. In other words, the address determines what the number stands for. For example, it may be an instruction to move the tool along the X axis, or to select a particular tool. It is illustrated in the following example:

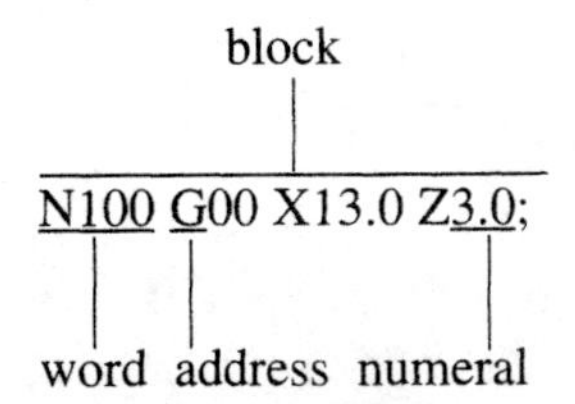

In the format for the computer word *X00005000,* the numeral 5 may mean that the move will be 0.5 inch. It is assumed that the decimal point is in the middle of the eight-digit number as follows: X0000.5000. This is an example for inch data input on a control with an eight-digit format. This format is a standard feature on today's controls. When using an eight-digit format, the maximum number entered would be 9999.9999. On most machines this value is not suitable because of the mechanical limits. This is especially true when using inch data input.

There are also controls with a seven-digit format. Then the maximum number entered would be 999.9999. Note that the number of decimal places remains the same, enabling the machine to work with the same preciseness when using either seven- or eight-digit format.

The zeroes before the first significant digit are called the *leading zeroes*. The zeroes behind the first significant digit are known as the *trailing zeroes*. A letter X determines what the number stands for. Thus, in the example, it would be a 0.5 inch move on the X axis.

Most controls allow suppressing the leading zeroes when entering data. This is known as **leading zero suppression.** When this method is used, the machine control reads the numbers from right to left, allowing zeroes to the left of the significant digit to be omitted. For example, Z1000 is entered for a 0.1 inch move on the Z axis, 10 is entered for an incremental offset adjustment of 0.001 inch, and F35 is entered for the feed rate of 0.0035 inch.

Some controls allow entering data without using the trailing zeroes. Consequently, it is called **trailing zero suppression.** The machine control reads the numbers from left to right, and zeroes to the right of the significant digit may be omitted. For example, Y00001 is entered for a 0.1 inch move on the Y axis. Note that the last 3 zeroes are omitted.

Most of the controls use one of the preceding formats, usually the leading zero suppression and decimal point format. Beginners on CNC controls should use the decimal point format if the machine has that capability. This is more natural to us, and, consequently, it is safer. Then, for instance, 0.5 or just .5 instructs a 0.5 inch move.

Note that the suppression technique depends on the particular control. Thus, one must know which format is acceptable when programming or program editing. Zero suppression and decimal point format are popular techniques because they shorten the coding when programming. Following are some additional notes about format:

1. On newer controls, the values with or without a decimal point can be specified together, such as in X4.55 Y3000 or Y1000 Z-1.3.
2. The values less than the least input increment are truncated. For instance, if Z1.23456 is specified, Z1.2345 is assumed in inch data input (4 decimals) and Z1.234 is assumed in metric data input (3 decimals).
3. The number of digits must not exceed the maximum number of digits allowed. When a number with a decimal point is input, it is converted into an integer of the least input increment, as in this example: 123456.7 is converted to 1234567000 (10 digits) in inch data, and 123456700 (9 digits) in metric data. In both cases the alarm will occur.
4. To enter negative numbers, use a minus sign, as in, X-5000 or Z-1.375. A plus sign is not necessary for positive numbers because the computer interprets any number without a sign as a positive number.

The End of Block Code

Each block of information in the program must end with an **end of block code,** which is a semicolon (;) in ISO, or EOB in the EIA code set. It tells the control about one action to be executed as one command. In the next block another action will be executed, and so on.

The end of block code allows you to tell the control what to do in each block. This is a very important matter since the program is always executed in blocks. The control reads and executes the information between two end of blocks. If there is no end of block between two commands in the program, the control considers these commands as one action. Figure 1–18 illustrates how dangerous this can be.

The instruction is given as X-1.0 Y1.5;. First, the tool is to be moved 1.0 inch from point A to point B in the minus X direction. Then, it is to be moved 1.5 inch from point B to point C in the plus Y direction, as per position (a). However, if the end of block is missing between these two commands, the control would interpret this as one move from point A to point C. The tool would be moved as per position (b). It would probably result in a scrapped part and tool damage. To achieve the expected result, the instruction has to be written with the end of block between these two commands:

```
X-1.0; Y1.5;
```

Experienced programmers write blocks in lines, no matter how short they are. This also makes the program easier to read. For example:

```
X-1.0;
Y1.5;
```

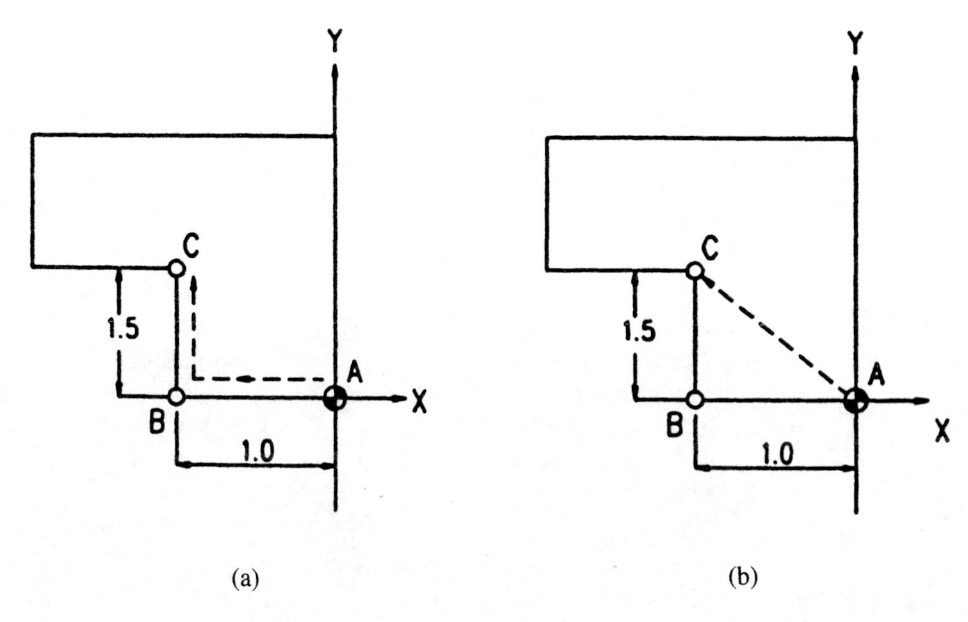

Figure 1–18 End of block is missing.

On older machines with small memory capacity, programmers can save space by entering the blocks one after another, as many as can fit on the line, but this is a poor practice and should be used only in extreme cases. Once again, programmers should take care not to forget to enter the end of block code between the programmed blocks.

The end of block code must appear on the machine monitor after each programmed instruction. When entering this code by manual data input on the machine, programmers and operators should use the EOB button or, on newer machines, the ENTER button. Office computers do not display the end of block code, but it appears on the machine monitor after the program loads. It will be shown at all the places where the programmer pressed the RETURN key.

The end of block code performs two functions, line feed and carriage return. It tells the machines to execute the previous information and to start reading the next data. Thus, it does no harm if there are several end of block codes one after another. They are sometimes left on purpose to separate groups of data in the program. It makes the program easier to read.

Types of CNC Codes

There are many codes in CNC language, and each code has its purpose. All of the codes do not have the same function on all machines. Many machines use codes that are unique to them. However, all CNC codes can fit into the following two categories:

1. Preparatory codes
2. Miscellaneous codes

Preparatory codes, as the name says, prepare the machine to treat the information in a distinct manner and to execute it. In essence, the preparatory function determines the mode of the system. These are all codes with the address G. For instance, if the programmer enters G20, inch input mode is set. It means that the system will use inch data input until it is changed to metric data input. Then, that particular G code will reset the system. On almost all controls, the codes G00, G01, G02, and G03 belong to the same group. For the other groups of G codes, consult the manual for the particular machine. On most machines there are four categories of G codes. They have the following purposes:

1. Selection of a measurement system (G20 and G21)
2. Selection of a movement system (G00, G01, G02, and G03)
3. Application the tool radius compensation (G40, G41, and G42)
4. Selection of preprogrammed sequences or canned cycles (G70, G71, G73, G81, G83, etc.)

More than one G code may be programmed in a block, such as G00 G80 G90, but they must be from the different groups. For instance, when the following line of code is programmed, G01 G00 X3.0;, the G00 code will be executed, producing rapid tool motion in the X axis, which may be unwanted. This also means that the codes from the same group cancel each other. If two G codes from the same group are programmed in one block, the last programmed will be executed.

Miscellaneous codes work like on/off switches for the functions they control. These are all codes with the address M. They control the auxiliary functions such as coolant on/off, spindle start/stop, and so on.

In principle, all codes are modal or nonmodal. **Modal codes** stay in effect until cancelled by another code in the same group. The control remembers modal codes. This gives the programmer an opportunity to save programming time and computer memory, which might be important when programming for machines that have a small storage capacity.

For instance, G01 is a modal code that produces linear interpolation. If the programmer wants to continue with linear motion, he or she does not have to program this code again because it is modal. This code would be cancelled if, for instance, the G02 code is programmed. Then, the linear interpolation would be changed to circular interpolation.

Nonmodal codes stay in effect only for the blocks that they are programmed. Afterwards, their function is turned off automatically. For instance, G04 is a nonmodal code to program a dwell. After one second, which is, say, the programmed dwell time in one particular case, this function is cancelled. To perform dwell in the next block, this code has to be reprogrammed. The control does not memorize nonmodal codes, so they are known as one shot codes.

All the M codes are modal, while the G codes can be either modal or nonmodal. Usually, only one M code is allowed per block. If the address has more than one meaning, the preparatory function G or miscellaneous function M will determine the exact meaning. For example, the address X can be specified to express the coordinate value, say X2500, or to command dwell, say X2500. In the first case, the G01 code is used, while in the second case, the programmer enters the G04 code to express what he or she wants to accomplish. For example:

G01 X2500; (Positioning to a 0.2500 diameter on the X axis in leading zeroes suppression format.)
G04 X2500; (Dwell for 2.5 seconds.)

A list of all CNC codes, modal and nonmodal, appears in Appendix A.

Summary

Numerical Control (NC) is the machining process in which the operations are executed automatically in sequences as specified by the program that contains the information for the tool movements. When Numerical Control is performed under computer supervision, it is called Computer Numerical Control (CNC).

NC technology has found many applications, including lathes and turning centers, milling machines and machining centers, punches, electrical discharge machines (EDM), flame cutters, grinders, and testing and inspection equipment.

There are many advantages of the CNC machines over conventional machines, such as the possibility of performing multiple operations on the same machine in one setup, a significantly reduced scrap rate, the ease of incorporating part design changes when CAD/CAM systems are used, the ease of performing quality assurance by spot check instead of checking all parts, and significantly increased production, just to name a few.

Generally speaking, any CNC machine tool consists of the following units: computers, control systems, drive motors, and tool changers. The control systems used on NC/CNC machines are open-loop and closed-loop. The open-loop control does not provide positioning feedback to the control unit, while in the closed-loop control this is monitored by the feedback device.

The majority of NC/CNC machine tools are equipped with automatic tool changers, such as magazines on machining centers and turrets on turning centers. They allow tool changing without the intervention of the operator. The tool changers may be equipped for either random or sequential selection. In random tool selection, the tools do not have to be loaded into the magazine or turret in the order in which they are called for in the program. The machine control knows where to find a particular tool. In sequential tool selection, the tools must be loaded in the exact order in which they are called for in the program; otherwise an unwanted tool may be loaded in the spindle.

CNC machine tools use two basic types of programming when producing a part: positioning or point-to-point and contouring or continuous-path. Positioning or point-to-point programming is used to move the tool from one point to another when programming operations such as drilling, tapping, boring, and reaming. After the operation is finished at one location, the tool travels to another. The main characteristic of positioning is that the tool is not in constant contact with the part. In contouring or continuous path programming, the tool is in constant contact with the part until the operation is finished. The tool is moved by the specified feed rate, not by rapid traverse, as in positioning. Milling and turning are the most common operations when contouring.

On NC/CNC machines there are reference points which must be thoroughly understood before attempting to program and operate such equipment. The machine zero point or the machine origin is a fixed point set by the machine tool builder. Any tool movement is measured from this point. The control always remembers the tool distance from the machine origin. A part zero point is the origin of the absolute coordinate system. It can be set at any point inside the machine's electronic grid system limits. Relocation of the part origin to another location is known as zero shifting. All the newer machines are designed to perform full zero shifting. The tool start point is the program origin. It is on a certain distance away from the part origin. Practically, a program origin may be described as the distance from the tool tip to the part origin when the tool is at the tool change station.

A NC/CNC machine uses a coding system. A frequently used coding system is the Binary-Coded Decimal, also known as the EIA code set. The newer coding system is the American Standard Code for Information Interchange, or ASCII. Because of its wide acceptance, it has became the International Standards Organization (ISO) code set.

CNC machines use a set of rules to enter, edit, receive, and output data. These rules are known as CNC syntax, programming format, or tape format. The format specifies the order and arrangement of information entered. If the control finds any discrepancy according to CNC syntax, a format error or an alarm message will occur. The most common CNC format is the word address format, which consists of an address (one of the letters from A to Z) followed by numerals. The address defines the meaning of the number that follows the address.

All CNC codes can fit into the following two main categories: preparatory codes and miscellaneous codes. Preparatory codes decide the mode of the system and execute the programmed information. They are all codes with the address G. Miscellaneous codes work like on/off switches for the functions they control. These are all codes with the address M. In principle, all codes are modal or nonmodal. Modal codes stay in effect until cancelled by another code in the same group. The control remembers modal codes. Nonmodal codes stay in effect only for the blocks that they are programmed. Afterwards, their function is automatically turned off. The programmer and operator must have a clear understanding of all the codes used on one particular CNC machine.

Key Terms

alternative current servomotors
automatic tool changer
closed-loop control system
Computer Numerical Control (CNC)
continuous-path programming
contouring
direct current servomotors
EIA code set
end of block code
ISO code set
leading zero suppression
machine origin
machine zero point
miscellaneous codes
modal codes
nonmodal codes
Numerical Control (NC)
open-loop control system
part origin
point-to-point programming
polar coordinate system
positioning
preparatory codes
program origin
random tool selection
rectangular coordinate system
sequential tool selection
stepper motors
trailing zero suppression
word address format

Self-Test

Answers are in Appendix E.

1. _______________ is any machining process in which the operations are executed automatically in sequences as specified by the program.
2. _______________ is Numerical Control performed under computer supervision.
3. _______________ convert a digital pulse, generated by the microcomputer unit (MCU), into a small step rotation.
4. _______________ are variable speed motors that rotate in response to the applied voltage.
5. _______________ are controlled by varying the voltage frequency to control the speed.
6. _______________ does not provide positioning feedback to the control unit.
7. _______________ is very accurate. Most of these have an automatic compensation for error, since the feedback device indicates the error and the control makes the necessary adjustment.
8. _______________ allow tool change without the intervention of the operator.
9. In _______________, the tools do not have to be loaded into the magazine or turret in the order in which they are called for in the program.
10. In _______________, the tools must be loaded in the exact order in which they are called for in the program.
11. _______________, or point-to-point programming, is used to move the tool from one point to another.
12. In _______________, the tool is in constant contact with the part until the operation is finished.
13. _______________ has two perpendicular axes in a two-dimensional surface, which is known as a plane.
14. On _______________, a point is defined by the radius vector and angle.
15. The _______________ is a fixed point set by the machine tool builder.
16. The _______________ is the origin of the absolute coordinate system.
17. The _______________ is the tool start point, which is related to the part origin and the machine origin.
18. The _______________ is a coding system developed by the Electronics Industries Association (EIA).
19. The _______________ is the ASCII code set accepted by the International Standards Organization (ISO).
20. _______________ is a technique that allows suppressing leading zeroes when entering data.
21. _______________ is a technique that allows suppressing trailing zeroes when entering data.
22. The _______________ is specified using a semicolon (;) in the ISO, or EOB in the EIA code set.

23. _______________ prepare the machine to treat the information in a distinct manner and to execute it.
24. _______________ work like on/off switches for the function they control.
25. _______________ stay in effect until cancelled by another code in the same group.
26. _______________ stay in effect only in the blocks for which they are programmed.

Relating the Concepts

No answers are suggested.

1. Define the terms NC and CNC.
2. List the applications of NC/CNC technology.
3. List the advantages and disadvantages of CNC machines.
4. What is the advantage of closed-loop control over open-loop control?
5. What is the advantage of automatic tool changers?
6. Differentiate between positioning and contouring.
7. List and describe the reference points on NC/CNC machine tools.
8. Why it is important that the machine allow full zero shifting?
9. List the coding systems used on NC/CNC machines.
10. Explain the importance of CNC syntax.
11. List the categories of CNC codes.
12. Differentiate between modal and nonmodal codes.

2 Programming Fundamentals

Key Concepts

Programming Modes

Programming on Diameter and Radius

Parameter Setting

Programming Functions

- Data Input
- Coordinate System Preset
- Tool and Tool Offset
- Spindle Control
- Feed Rate Control
- Move Commands
- Dwell
- Program Control
- Subprogram Control

Programming Efficiently

Basic Program Structure

Program Loading

Program Proving

Programming Modes

The two basic methods of positioning a tool for movement from one location to another are absolute and incremental, also known as relative positioning. Positioning can be done in either manual or automatic mode.

The terms absolute and incremental positioning are interchangeable with the terms absolute and incremental programming. In **absolute programming,** all measurements are made from the part origin established by the programmer and set up by the operator. Any programmed coordinate has the absolute value in respect to the absolute coordinate system zero point. The machine control uses the part origin as the reference point in order to position the tool during program execution.

The sign of each coordinate depends on where the tool is moving according to quadrants. On the lathe, the sign for all the X and Z coordinates in the first quadrant is positive. In the same way, the machining center X and Y coordinates are positive in the first quadrant. In the second quadrant, the lathe X and machining center Y coordinates are positive, while the lathe Z and machining center X coordinates are negative, and so on, according to quadrants.

In **incremental programming,** the tool movement is not measured from the part origin; it is measured from the last tool position. The programmed movement is based on the change in position between two successive points. The coordinate value is always incremented according to the preceding tool location. The programmer enters the relative distance between the current location and the next programmed point. The sign of each coordinate depends on where the tool is moving with respect to the machine axis. For example, to move the tool upward on the machining center Z axis, he or she enters a positive sign. To move the tool downward, he or she enters a negative sign.

When programming for the lathe, use the X and Z addresses for absolute programming, and U and W for incremental programming. Switching from one mode to another is easy; just specify the address and its value. Figure 2–1 shows the program for the possible tool movements for the part.

To move the tool from point A to point B in absolute, the command block is programmed as:

```
Z-2.5;
```

To move the tool on the same path in incremental, the command block would be programmed as:

```
W-1.0;
```

In the first case, the tool is moved to point B in relation to the part origin. In the second case, the tool is moved to point B in relation to point A. In both cases, the sign is negative. Also, the tool is moved the same amount. This means that the amount of the tool motion is the same in either absolute or incremental programming.

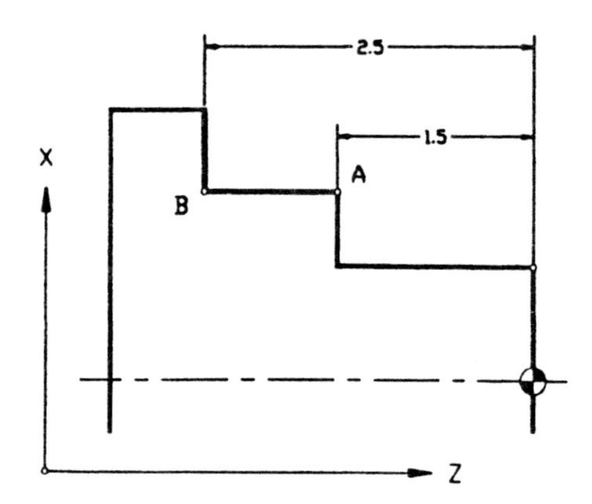

Figure 2–1 Absolute and incremental programming on the lathe.

Absolute and incremental programming may be combined in the same program, even in the same block. The following is a valid command to, say, cut the chamfer on the lathe:

G01 X3.0 W-0.5;

When programming for the machining center, the programmer must specify the G90 code in absolute programming or the G91 code in incremental programming. For example:

G90 X-2.5; (Switching from incremental to absolute.)
G91 X-1.0; (Switching from absolute to incremental.)

To further illustrate incremental programming, consider the following case. The tool is cutting 3.0 inches below the top part face where the part origin is set. To continue cutting 1.0 inch deeper in incremental, the programmer would enter:

G91 Z-1.0; (The move is in the minus Z direction.)

The tool is cutting to the depth of 4.0 inches. The incremental move from this location to the position of 2.0 inches from the part origin would be programmed as:

G91 Z2.0; (Note that the move is in the plus Z direction.)

Assuming that there was no change in programming mode between the two previous commands, the G91 code does not have to be entered since it is a modal code. However, it does no harm if specified.

The same task may be programmed using absolute programming or combining the absolute and incremental modes, as in the following examples:

Example 1 Using absolute programming.

G90 Z-4.0; (The tool is cutting to the depth of 4.0 inches.)
Z-2.0; (The tool is moving to the depth of 2.0 inches.)

Example 2 Combining absolute and incremental programming.

G90 Z-4.0; (The tool is cutting to the depth of 4.0 inches in absolute.)
G91 Z2.0; (The tool is moving to the depth of 2.0 inches in incremental. Note that the move is in the plus Z direction.)

There are advantages and disadvantages to both the absolute and incremental programming modes. For instance, in absolute programming the programmer always knows where the tool is in relation to the part origin. When using incremental programming, it is easier to comprehend what the tool is doing. In absolute, it is easy to return the tool to the start position just by entering the same values as in the coordinate system preset block. To return the tool in incremental, the programmer has to make the calculation in order to know how far the tool has moved from the start position. For one change in the program written in absolute, one block has to be changed, while in incremental, two blocks have to be changed. When programming in absolute, one mistake in the program scraps one part size, while in incremental, the mistakes are chained, causing several part sizes to be scrapped.

Which programming mode to use depends on how the part is dimensioned and what the programmer feels is more convenient. In general, absolute programming is normally used, but one can start programming in one mode, say absolute, then switch to incremental because it is easier to calculate incremental values from a blueprint. The programmer can switch back to absolute at any time.

Programming on Diameter and Radius

The CNC lathe programming may also be distinguished as **programming on diameter** and **programming on radius.** Which type of programming will be used is set by the parameter. Older machines accept only one type of programming, usually programming on diameter, while newer machines use either one. Figure 2–2 illustrates several blocks programmed to move a tool from point A to point E.

	DIA PROGRAMMING	RADIUS PROGRAMMING
Point A	X1.0 Z0;	X0.5 Z0;
Point B	Z-0.5;	Z-0.5;
Point C	X1.5	X0.75;
Point D	Z-1.5;	Z-1.5;
Point E	X2.0;	X1.0

In practice, lathe programming on diameter is used more because the part drawings for lathe work are mainly dimensioned by diameter. Thus, they can be directly substituted for the X coordinates entered in the program.

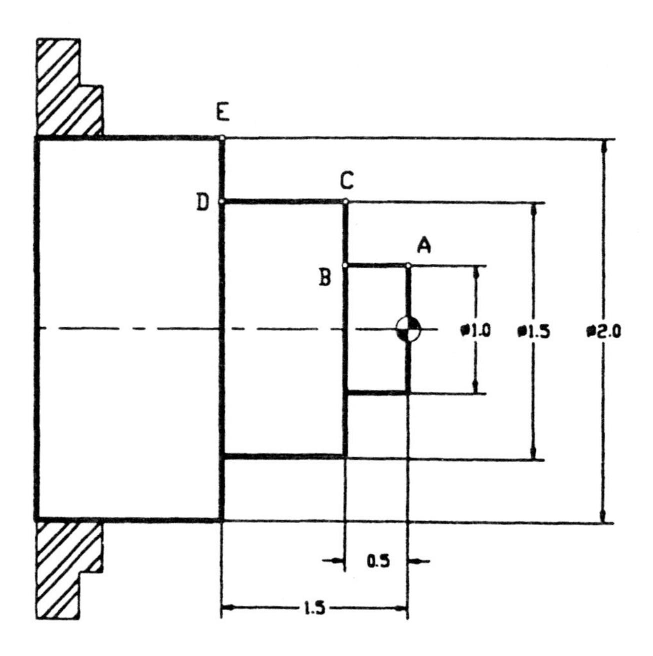

Figure 2–2 Programming on diameter and radius.

Parameter Setting

Parameters are selected instructions by which the machine is programmed, set up, and operated. These instructions might be called defaults, or first choices when the machine is turned on. For example, if a programmer more often uses inch data input, he or she can set the particular parameter to invoke the G20 code as the first choice. This means that this code does not have to be programmed since it is selected automatically. However, the programmer can override the parameter setting by programming the G21 code, selecting metric data input.

There are many machine functions that can be affected by the parameter setting, including the feed rate mode, the type of programming, the amount of rapid traverse rate for each axis, whether the dry run is effective or not, and the data input mode selection. Older machines have fewer parameters that can be set or changed by the programmer or operator. Newer machines are much more sophisticated and have hundreds of parameters. In general, the more parameters the machine has, the more flexible and sophisticated it is.

When we look at the Parameter screen, reached by pressing the PARAMETER button on the machine control, we may see many numbers similar to this one:

NO 0001　　01011000

This means that on this particular machine, parameter number 0001 has the following setting, an eight-digit number: 01011000. The ones and zeroes are not the machine codes. They are just the switches used by the control to determine how the particular parameter has been set. For instance, when the first digit is 0, it may mean that the program tape is not checked for errors vertically. If the programmer wants the tape checked vertically for errors, he or she changes 0 to 1. The second digit may be assigned to the code used for data output. For instance, if the programmer sets 0, the EIA format is used; if he or she sets 1, the ISO format is used, and so on for each digit in the eight-digit number.

The parameters should be of interest for the programmer and operator in order to improve their understanding of the CNC equipment. The machine manual of one particular machine lists all parameters used, as the meaning of the switches 0 and 1, and some other switches which are also sometimes used.

The use of the parameter setting in the machine shop practice may be explained in the following way. When programming for the lathe, you may decide that the feed rate per time should be used instead of the feed rate per revolution. After checking the machine manual, you may find the place where it is described under the parameter number, such as 1130. For example, the manual may say that the fifth digit sets the feed rate:

1 : Feed rate per minute. The G98 code is set.
0 : Feed per revolution. The G99 code is set.

Then you list the parameters through the screen in order to find parameter number 1130. After reaching it, you change the fifth digit from a 0 to a 1 according to protocol. Consequently, the feed rate per time will be set as the first choice, either in manual or memory mode. The control knows this after checking parameter number 1130, where it finds that the fifth digit is 1.

The parameters are set when the machine tool builder representative sets up the machine, and only a few are changed once in a while. Sometimes the program just does not work properly, although everything looks good. It is then wise to check the machine manual to see how the parameters should be set, because some of them might have been changed. Keep in mind that parameter setting is the area where CNC machines differ the most.

Programming Functions

Programming functions include the machine G and M codes and some switches. Many of them look simple, but they are not. For instance, the **slash code** (/) is a very helpful programming function that may be used to solve the most difficult programming tasks. However, it is often described as a simple block delete or optional skip code.

In order to understand what to program and when, programmers and operators should know some basic functions found on all CNC machine tools, which are discussed in this chapter. The more advanced programming functions are presented in the chapters that follow.

Data Input

Almost all CNC machines use **inch** and **metric data input.** The parameter setting distinguishes which data input is in use. If data are usually entered in inches, then the parameter should be set as such, and it becomes the default or the first choice. When conversion from one data input to another is needed, use the following conversion:

1 inch = 25.4 mm

For instance, convert 3.0 inches into metric units by multiplication:

3.0 • 25.4 = 76.2 mm

To convert metric units into inch units, divide, as in the following example to convert 100 mm:

100 / 25.4 = 3.937 inches

Unit conversion takes time. Thus, it is more efficient to program using the units in which the parts are shown on the blueprint.

When switching from one data input mode to another, the offset values have to be carefully checked. For example, 0.1 mm offset in metric data input becomes 0.1 inch in inch data input. This makes a considerable difference. Remember that when switching from one data input mode to another, the unit system of the following items may be changed:

1. Feed rate
2. Positional command
3. Offset value
4. Unit of scale for manual pulse generator
5. Some parameters

It is not acceptable to use both data input modes at the same time or to switch from one to another in the same program.

When using the G20 or G21 codes, the code must be programmed in a single block and usually on the first line in the program, but at least before the coordinate system preset.

Coordinate System Preset

Each tool must have its **coordinate system preset** block in order to establish the program origin. For that purpose, the G50 function on the lathe and G92 function on the machining center are usually used.

When reading the coordinate system preset block, the control assumes that the values entered describe the tool distance from the part origin in respect to each machine axis. For example:

G50 X10.0 Z2.0; (Lathe instruction format.)
G92 X15.0 Y8.0 Z10.0; (Machining center instruction format.)

According to these instructions, the distance from the lathe tool tip to the part origin is 10.0 inches on the X axis and 2.0 inches on the Z axis. The distance from the machining center tool tip to the part origin is 15 inches on the X axis, 8 inches on the Y axis, and 10 inches on the Z axis.

The values for the G50 or G92 block have to be in absolute, not incremental, programming. Consequently, the X and Z on the lathe and the X, Y, and Z on the machining center are used to preset the coordinate system. This is established by the machine tool builder.

When programming more than one tool, either the same or different values may be used in the coordinate system preset block. The operator applies different techniques for set up. Sometimes it is necessary to correct the values entered in the the coordinate system preset block. For that purpose, a feature called offset is used. This feature will be discussed fully in the following chapter.

On the lathe, the G50 function is also used to set the maximum spindle RPM (revolutions per minute), which will not be exceeded even if a higher speed is entered later in the program. This is known as the maximum cutting speed or **spindle speed limit.** For example, in the lathe programming instruction G50 S1500, the speed limit is 1,500 RPM.

The spindle speed limit may be programmed in a later block after the coordinate system preset block, as in the following series of instructions:

G50 X10.0 Z3.0 M42; (Coordinate system preset, high range.)
G00 T0100; (Tool and tool offset selection.)
G50 S1000; (Speed limit.)
G00 X5.0 Z0.2; (Rapid to position.)

In general, any S code preceded by the G50 code is assumed to be the maximum spindle speed setting code. The ordinary spindle speed selection for the individual cuts cannot be made in the G50 block. For instance, the instruction G50 S900 tells the control that the speed limit is 900 RPM, but it cannot turn the spindle using the programmed data. To turn on the spindle, the programmer must enter the following instruction: S900. Then, he or she would add the M03 or M04 code if needed.

Programming the G50 on the lathe and G92 on the machining center is not the only way of presetting the coordinate system. On some newer machines the G00 shift or G54 through G59 functions may be used. They are very safe features which reflect the newer tendencies in designing CNC machine tools. However, you cannot fully understand these functions until you understand the standard G50 and G92 functions.

Tool and Tool Offset

The T function is used to call the particular tool and **tool offset** in the program. The tool offset is used to correct the values entered in the coordinate system preset block. This can be done quickly on the machine without actually changing the values in the program.

Using the tool offsets, it is easy to set up the tools and to make adjustments in part size. Tool offset is one of the greatest advantages of the CNC machine over the conventional machine. (This feature will be fully discussed in the next chapter.)

On the lathe, calling and cancelling the tool and tool offset is expressed with the address T and a two- or four-digit number, which can be set by the parameter. The following is the lathe instruction format:

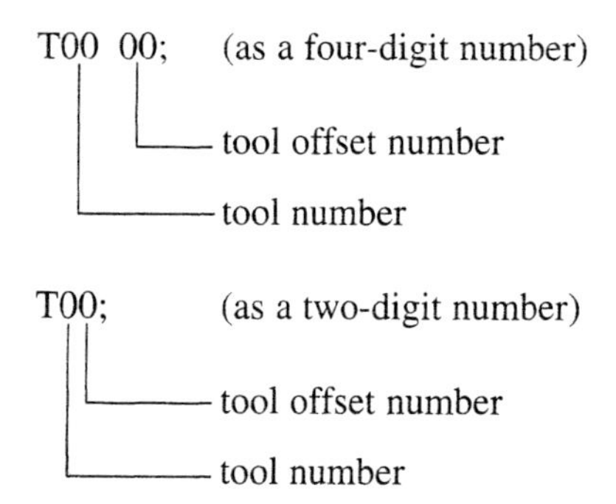

On the machining center, the T address is used to call the tool number. The tool offset is usually called using the H address. The following is the machining center instruction format:

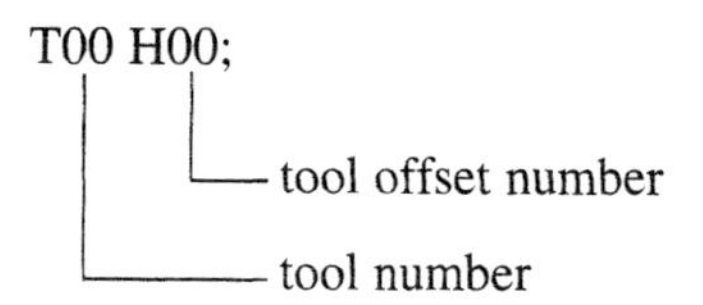

The tool numbers usually range from 01 to 12 on the lathe and from 01 to 32 on the machining center. The tool offset numbers usually range from 01 to 99 in groups of 8 or 16. The numbers used depend on the machine tool builder.

When calling the offset on the machining center, the G43 code is usually used in order to apply the offset by moving the tool toward the part. On the same machine, the tool offset is usually cancelled by the G49 code. The tool and tool offset functions are used together with the G00 code. For example:

G00 T0505; (Tool number 5 and offset number 5 call on the lathe.)
G00 T0500; (Tool offset cancel on the lathe.)
G00 G43 H10; (Offset number 10 call on the machining center.)
G00 G49; (Tool offset cancel on the machining center.)

On newer machines, tool offset can be cancelled by simply entering T0 instead of the tool offset number. It can be used either on the lathe or machining center. This makes the programming job easier.

Spindle Control

The spindle speed is programmed by the S code and a four-digit number, such as S1350. The leading zeroes do not have to be entered, thus S550 can be programmed instead of S0550. On some older machining centers, the speed is expressed indirectly. For instance, if 355 spindle RPM is desired, the programmer enters S30 according to a chart supplied with the machine.

There are two spindle speed modes used on CNC machines:

1. Revolutions per minute (RPM)
2. Constant surface speed

The spindle speed in revolutions per minute is also known as **constant RPM** or direct RPM. The change in the tool position does not affect the RPM commanded. It means that the spindle RPM will remain constant until another RPM is programmed.

On the majority of machining centers, constant RPM is the only spindle speed mode used. Thus, there is no need for a special code when programming for these machines. On most of the lathes, the constant RPM is programmed using the G97 code.

Constant surface speed is almost exclusively used on the lathes. The RPM changes according to the diameter being cut: the smaller the diameter, the more RPM is achieved; the bigger the diameter, the less RPM is commanded. This is changed automatically by the machine speed control unit while the tool is changing positions. This is the reason this spindle speed mode is also known as the diameter speed.

Using the constant surface speed, it is possible to get very smooth finishing when facing or shoulder cutting. This cannot be achieved on conventional machines that use constant RPM.

The majority of machining centers do not use constant surface speed. On lathes, constant surface speed is selected using the G96 code when programming operations such as facing, turning, or boring. Then, it is enough to choose the cutting speed for the particular material, and the control will take care of the spindle RPM at any point of the tool path. If the programmed X axis coordinate is negative, the control uses the absolute X value to calculate the RPM. When the machine is in the Machine lock status, the RPM is calculated according to the changes in the X coordinates, although the tools do not move. Constant surface speed is expressed in ft./min. (feet/minute) in inch data input or m/min. (meters/minute) in metric data input.

To prevent the spindle RPM from exceeding the permitted spindle RPM set by the machine tool builder, the programmer uses the G50 function to set the maximum spindle RPM. For small chuck machines the higher speed limit, such as 2,000 RPM, is set by the machine tool builder. For bigger and heavier chucks, a much lower speed limit is set, sometimes just several hundred RPM. To make sure that the speed limit will not be exceeded, check the machine manual when programming or editing the program for a particular machine.

Constant surface speed control is effective in the thread cutting mode, but it is better to cancel it when cutting taper or face thread in order achieve the correct thread pitch.

On the lathe, the G96 or G97 function is combined with the S code, as in the following examples:

G96 S350 (Constant surface speed of 350 ft./min.)
G97 S350 (350 RPM)

As seen, the G96 and G97 codes are used to express the spindle speed mode, but not the spindle speed amount. For that purpose, the S function is used.

The following is a relation between the constant surface speed and constant RPM:

$$\text{Inch RPM} = \frac{V \bullet 3.82}{D}$$

$$\text{Metric RPM} = \frac{1000 \bullet V}{D \bullet \pi}$$

V Velocity or cutting speed
π Constant = 3.14
D Diameter of part or tool

For example, the cutting speed for 1020 steel is 550 ft./min. What is the RPM to machine a 7.0 inch part diameter on the lathe?

$$\text{Solution: RPM} = \frac{550 \times 3.82}{7.0} = 300$$

As seen from this example, when cutting a 7.0 inch part diameter on 1020 steel, the programmer can enter G96 S550; or G97 S300;. However, the G96 function enables the proper RPM for any diameter being cut.

Both of these codes (G96 and G97) are modal, and they cancel each other. When the lathe is turned on, the G97 mode is usually chosen automatically by the parameter settings. Then, cutting is performed by direct RPM. When entering the spindle speed in the lathe program, it is necessary to specify whether to use the constant surface speed control or constant RPM.

To start and stop the spindle, either on the lathe or the machining center, the following codes are used:

M03 Turns the spindle in a clockwise direction viewing from the headstock toward the tailstock on the lathe or viewing toward the table on the machining center
M04 Turns the spindle in a counterclockwise direction viewing from the headstock toward the tailstock on the lathe or viewing toward the table on the machining center
M05 Stops the spindle rotation

On some machines, the spindle has to be stopped before the direction is changed. This means that the M05 code has to be programmed between the M03 and M04 codes. It is set by the machine tool builder.

In order to use the spindle speed function efficiently, it is important to know the RPM limits for a slow (M41) and a high (M42) gear range on one particular machine. For instance, it is wrong to program a high range for heavy cutting conditions because heavy forces could stop the spindle. In this case, a low range should be programmed. It is also not recommended to use a low range for light jobs since the programmed speed may not be reached. For an illustration, assume that the maximum RPM for a low range is 450. Even if it is programmed 650 RPM for one particular job, the machine locks the speed at 450 RPM.

On some machines there are one or more subranges. On most machines, the spindle does not have to be stopped to change the range.

Feed Rate Control

Cutting operations may be programmed using two basic feed rate modes:

1. Feed rate per spindle revolution (G99)
2. Feed rate per time (G98)

The **feed rate per spindle revolution** depends on the RPM programmed. The specified feed rate of 0.01 inch means that the tool will move 0.01 inch for each revolution of the spindle. For instance, if the spindle is turning at 100 RPM, and 0.01 inch feed rate is specified, the tool will advance 1.0 inch per minute (0.01 • 100 = 1.0). Although the feed rate remains constant as programmed, the tool advances more or less, depending on the increase or decrease in the RPM. This type of feed rate is usually used on lathes.

The **feed rate per time** depends on the feed rate specified. The tool will advance for the amount of the feed rate programmed regardless of the spindle revolution. For example, when the programmed feed rate is 10 inches per minute, the tool will move for 10 inches in one minute, regardless of whether the spindle is turning fast or slow. The feed rate per time is expressed in inch/min. (inches/minute) in inch data input, or m/min. (meters/minute) in metric data input. It may also be programmed in deg./min. (degrees/minute) when using fourth or fifth machine axes. This type of feed rate is usually used on machining centers.

On the lathe, the feed rate per revolution is used almost exclusively. It may be programmed by the G99 or G95 code, depending on the make of the machine. The feed rate per time is also sometimes used. For example, it can be programmed in conjunction with dwell while some actions are performed, such as pulling bar materials by a pull-out finger.

On the machining center, the feed rate per time is normally used, but this can be changed to feed rate per spindle revolution if desired. The machine manual should be checked for the respective G codes.

The following are some examples of programming formats for the feed rate:

Example 1 Using decimal point format.

G98 F5.0; (Feed rate of 5 inch/min.)
G99 F0.015; (Feed rate of 0.015 inch/rev.)
G98 A0 F60.0; (Feed rate of 60 deg./min. The A address tells that the unit is deg./min. Without it, the feed rate would be 60 inch/min. In the first case, the rotary axis will rotate 60 deg./min., which is a slow motion. In the second case, the tool is moving fast. Thus, care is needed when programming the feed rate, especially when programming the feed rate in deg./min.

Example 2 Using leading zeroes suppression format.

G98 F1000; (Feed rate of 0.1 inch/min.)
G99 F25; (Feed rate of 0.0025 inch/rev.)

The amount of the feed rate specified can be changed during the program execution. For this purpose, the Feed override switch on the operator panel may be used. The change can be made in increments of 10% in a range from 0% to 200%.

Whether the programmer uses the feed rate per time or per spindle revolution, it can be set by the parameter setting. Then, it is in effect automatically when the power is turned on. As mentioned earlier, the particular G code is programmed when desired to override the parameter setting.

The G codes for the feed rate are modal and have to be commanded in a single block. They are used to express the feed rate mode in the program but not the feed rate amount. For that purpose the F code is used.

Usually, the G codes for feed rate mode are not seen in the program since one of them is set by parameter. Then only the feed rate amount is programmed, such as F1.0;. But one must be certain whether the machine is set to use the feed rate per time or per revolution. The feed rate of 1.0 inch is not unusual if it is feed rate per time, but if it is feed rate per revolution, then it could be a very excessive feed rate. For instance, if turning the part at a rate of 500 RPM using a 1.0 inch/rev., the tool would travel at 50.0 inch/min. (500 RPM • 1.0). This feed rate would damage the insert as soon as it comes into contact with the part.

Move Commands

The machine tool can be positioned in rapid mode using the G00 function or in cutting mode using the G01, G02, or G03 codes. The G00 and G01 codes are used in almost any program, while the programmer uses the G02 and G03 codes if an arc is to be cut. Each of these codes is modal, so once programmed they stay in effect until cancelled.

The G00 function is normally used for positioning when the tool is approaching or leaving the part. It is also used to call or cancel the tool and tool offset, as well as to cancel a canned cycle. It is a good practice to program this code at the beginning of the program, especially on the machining center. This will ensure that all canned cycles are cancelled and all registers are set to the initial state. Following are some examples using the G00 code:

Example 1 G00 G90 X3.0 Y-1.25; (Rapid to position in absolute programming on the machining center.)

Example 2 G00 U0.2 W0.1; (Rapid positioning in incremental programming on the lathe.)

Example 3 G00 G80; (Canned cycle cancel on the machining center.)

Example 4 G00 T0505; (Tool and tool offset call on the lathe.)

The magnitude of rapid traverse depends on a particular parameter, which is set by a binary number. Thus, there is no need to program it. The rapid rate can be decreased manually from 100 to 50 and 25% using the rapid override switch. When the switch is on 0, it becomes the constant feed rate set by the parameter.

When programming this function on the lathe, the programmer should take care that the tool does not interfere with the part or machine, because both slides usually do not move by the same rapid rate. When the slides are programmed to move simultaneously, they move the tool diagonally (Figure 2–3).

The G01 function is called **linear interpolation,** *linear cutting,* or *straight cutting.* It is used to cut segments of a straight line between two points. It can also be used to cut approximate curves of an ellipse, since there is no programming function to cut an ellipse in one command.

In this cutting mode, the tool is moved from point to point by the programmed feed rate. This function cuts a straight line, taper, chamfer, and groove, but when needed, a radius, arc, and special curves can be cut too. Also, a cut-off can be performed. When using this code, the programmer must specify the feed rate. Following are some examples of programming the G01 function:

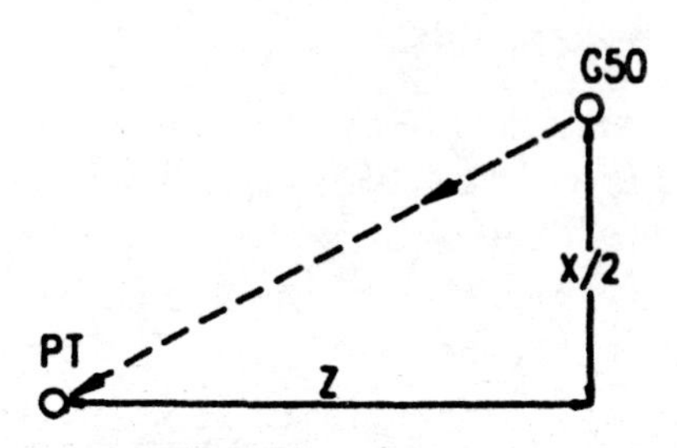

Figure 2–3 Simultaneous rapid motion.

Example 1 G01 X0 F0.01; (Cutting along the X axis in absolute programming on the lathe.)

Example 2 G01 U0.5 W-0.25 F0.008; (Cutting a 0.25 inch chamfer in incremental programming on the lathe.)

Example 3 G91 G1 X1.0 Y1.0 F6.0; (Cutting a 1.0 inch bevel on the machining center. Note that the leading zero is omitted and G1 is used instead of G01.)

If the feed rate is not specified in the block where this code is first programmed, most of the machines will generate an alarm. Afterwards, if the feed rate is not programmed, the last specified feed rate will be used by the control.

The G02/G03 function is used for **circular interpolation,** in which a tool cuts an arc, arc segment, or even a full circle. The tool is moved on the two axes simultaneously. The arc can be cut in a clockwise (G02) or counterclockwise (G03) direction, as shown in Figure 2–4.

When programming circular interpolation, it is necessary to place the tool at a known start point on the arc. It may be either positioning to a new location for cutting or a continued movement from the previously programmed position. In an arc cutting block, the tool start point is not programmed, only its ending position. The following three pieces of information are required to program an arc in circular interpolation:

1. Cutting direction, G02 or G03
2. Arc end point coordinates
3. Arc center displacement

The cutting direction, G02 or G03, tells the control in which direction to move the tool. On machines with the left-hand coordinate system, these directions are interchanged. G02 becomes counterclockwise and G03 become clockwise.

Figure 2–4 Circular interpolation.

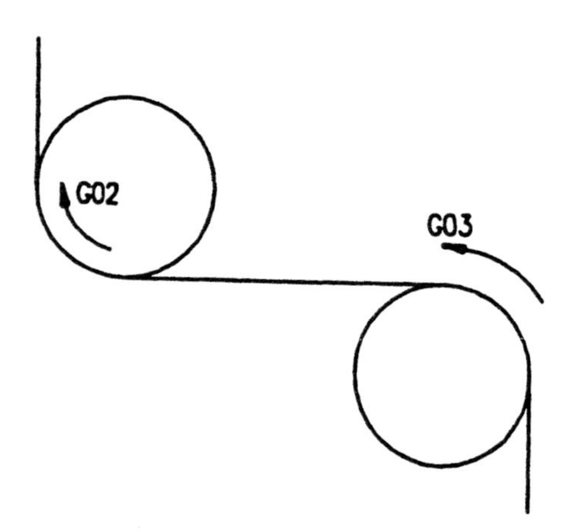

The arc end point coordinates, which tell the control where to move the tool from its present position, may be programmed in absolute or incremental programming. In absolute programming, the sign depends on the move direction according to quadrants; in incremental programming, the sign depends on the machine origin.

The **arc center displacement** is the incremental distance from the arc start point to the arc center. This distance is represented by the addresses I and K on the lathe and I, J, and K on the machining center. The sign for the I, J, or K address depends on the move direction according to the machine axes, either in absolute or incremental programming.

The end point of the arc is not modal and must be programmed for each arc. When two arcs are programmed in successive blocks, the control assumes that the programmed end point for the first arc is the start point for the next arc. Usually, I, J, and K are nonmodal and are programmed using the incremental values, either on the lathe or machining center.

On the lathe, the I address represents the incremental distance from the arc start point X coordinate and arc center. The K address represents the incremental distance from the arc start point Z coordinate and arc center (Figure 2–5).

On the machining center, the I address represents the incremental distance from the arc start point X coordinate and arc center. The J address represents the incremental distance from the arc start point Y coordinate and arc center. Because of the possibility of 3-D machining, there is an additional arc designation on the machining center for the K address. It is an incremental distance from the arc start point Z coordinate and the arc center (Figure 2–6).

If the previously specified feed rate is satisfactory, it does not have to be programmed in the arc cutting block. Figure 2–7 illustrates how to program the I and K values on the lathe.

When programming the tool to move from PT1 to PT2, the programmer gives the following instruction, ensuring that a 90-degree radius in counterclockwise direction is cut.

G03 X3.0 Z-0.25 I0 K-0.25 F0.009;

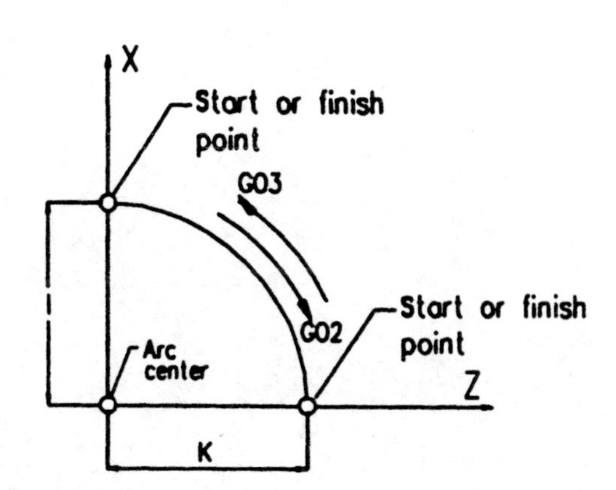

Figure 2–5 Data for arc cutting on the lathe.

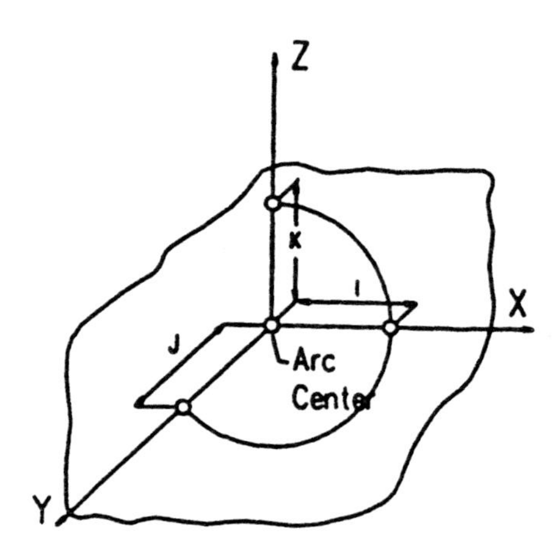

Figure 2–6 Data for arc cutting on the machining center.

X3.0 and Z-0.25 are the arc end point coordinates. The sign is minus on the Z address, because the tool is moving in the second quadrant. Accordingly, the sign for the X address is positive. The I address has zero value because PT1 (where the tool starts cutting) has the same X coordinate as the arc center. There is no displacement; thus I0 is programmed. It can also be omitted, and the control will assume that it has zero value. The K address has 0.25 value, because the tool start point is 0.25 inch from the arc center in the Z axis. The sign is negative since the tool is moving in a negative Z direction (away from the machine zero). The same task can be programmed in incremental mode using the W address:

G03 X3.0 W-0.25 I0 K-0.25 F0.009;

If the tool is programmed to move from PT2 to PT1 in a clockwise direction, the command block would be written as:

G02 X2.5 Z0 I-0.25 K0 F0.01;

Notice that now the K address is zero because the tool start point Z coordinate and the arc center Z coordinate are in the same line, meaning that they are the

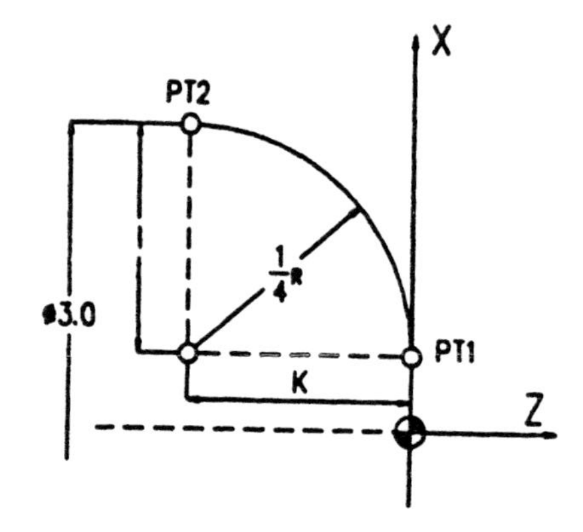

Figure 2–7 Calculating the values for I and K on the lathe.

same. Thus, there is no displacement and K0 is programmed (or it may be omitted). In the X axis, the tool start point is 0.25 inch from the arc center. The sign is minus because the tool is moving in minus X direction. Therefore, I-0.25 is programmed. Notice that the I value is not entered as a double value, which is the case with the X coordinates in diameter programming.

The same arc cutting can be accomplished in incremental mode using the W address instead of the Z address:

```
G02 X2.5 W0.25 I-0.25 K0 F0.01;
```

Note that when moving the tool from PT2 to PT1, the amount of tool motion is the same, but the values entered for incremental and absolute programming are not the same. To reach PT1 from PT2, Z0 is programmed in absolute, and W0.25 in incremental. Also notice that W is programmed as a positive value. This is because the tool is moving in a plus Z direction (toward the machine origin).

For arcs of 90 degrees, it is enough to program one of the I, J, or K addresses; the control will know where the arc center is. When an arc other than 90 degrees is cut, two of these addresses must be programmed in order to describe the position of the arc center in one plane (Figure 2–8).

From the triangle 0-PT1-M in Figure 2–8, the values for I and J are calculated as follows:

$$I = R \bullet \cos 30 = 1.0 \bullet \cos 30 = 1.0 \bullet 0.866 = 0.866$$
$$J = R \bullet \sin 30 = 1.0 \bullet \sin 30 = 1.0 \bullet 0.5 = 0.5$$

According to the same figure, to move the tool from PT1 to PT2 on the machining center, the instruction block would look like this:

```
G03 X0 Y1.0 I-0.866 J-0.5;
```

Note that in the previous example, both the I and J addresses have a negative sign.

Newer machines allow using the R address for circular interpolation. This eliminates the need for calculation of displacement values, since this function is

Figure 2–8 Programming an arc other than 90 degrees.

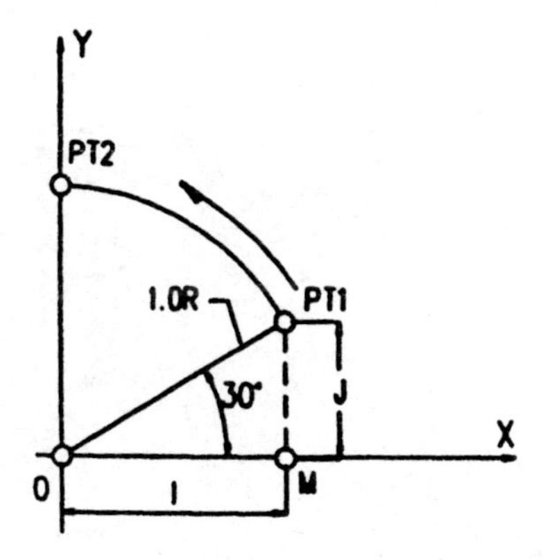

performed by the control. To move the tool from PT1 to PT2 using the R address, the programmer would enter the command block as:

G03 X0 Y1.0 R1.0;

On older controls, arcs can be cut in only one quadrant. To cut an arc in more than one quadrant, the programmer must enter two or more blocks. It is not necessary to repeat the G02 or G03 codes because they are modal. Also, the control assumes that the end point of each partial arc is the start point for the new arc (Figure 2–9).

To move the tool from PT1 to PT2 in Figure 2–9, two blocks must be programmed: the instruction to cut from PT1 to point M is placed in the first block, and the instruction to cut from M to PT2 is placed in the second block:

N50 G02 X0 Y1.0 I0.707 J-0.707; (First quadrant.)
N55 X0.707 Y0.707 J-1.0; (Second quadrant.)

All newer machines, either lathes or machining centers, allow cutting an arc in more than one quadrant in one programmed block. This is known as **multiquadrant circular interpolation.** On these machines, it is easy to cut the arc directly from PT1 to PT2 by entering the end point coordinates and the arc center displacement values:

N50 X0.707 Y0.707 I0.707 J-0.707;

To move the tool from PT1 to PT2 when using the R address, the instruction block would read:

N50 X0.707 Y0.707 R1.0;

Using the G02 or G03 function, even a full circle can be cut on the machining center. This feature is very useful on the machining center when the proper size drill, reamer, or end mill is not available (Figure 2–10).

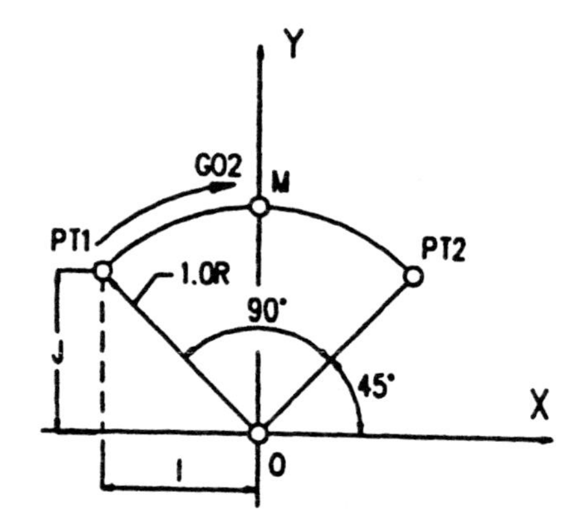

Figure 2–9 Cutting an arc placed in more than one quadrant.

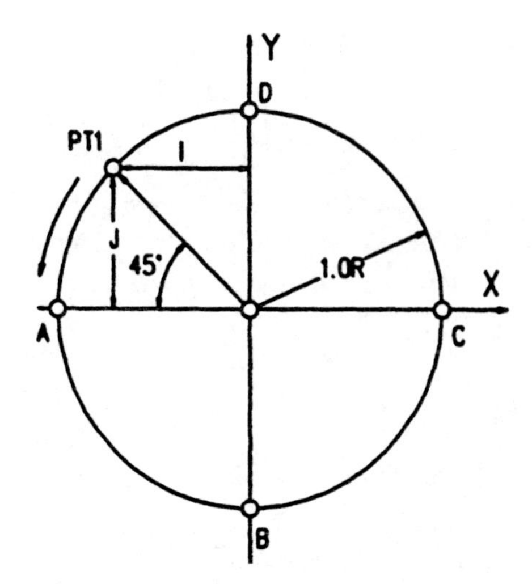

Figure 2–10 Cutting a 360-degree arc.

On machines without multiquadrant interpolation, cutting a full circle is programmed in several blocks. As in Figure 2–10, the program segment to move the tool from PT1 all the way back to PT1 in a CCW direction is:

```
G03 X-1.0 I0.707 J-0.707; (To point A.)
Y-1.0 I1.0; (To point B.)
X1.0 J1.0; (To point C.)
Y1.0 I-1.0; (To point D.)
X-0.707 Y0.707 J-1.0; (To PT1.)
```

On a machine with multiquadrant circular interpolation, this task is programmed in only one block. The same X and Y coordinates are specified, or, if they are omitted, it is assumed that they are the same coordinates as at the start point:

```
G03 I0.707 J-0.707;
```

Notice that when cutting a full circle using multiquadrant interpolation, the R address is not used. This is because the control does not have enough information about the actual tool position. Thus the I and J addresses must be programmed to cut a full circle in one instruction.

Circular interpolation can be programmed for any plane of the machining center coordinate system. A specific plane is selected by the codes G17 (X–Y plane), G18 (X–Z plane), or G19 (Y–Z plane) (Figure 2–11).

Plane selection is always made in the preceding block or in the block where a specific cutting instruction is programmed. For example: G18 G02 X1.0 Z0.5 I0.5;

If the plane is not specified, it might be assumed that the arc is to be cut in the machine basic plane (G17), which is parallel to the machine table. When programming for the lathe, it is not necessary to select the plane, because there is only one plane.

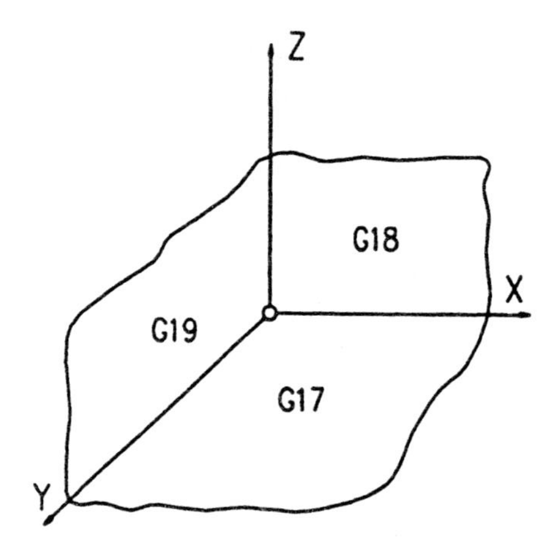

Figure 2–11 Planes of the machining center coordinate system.

Sometimes the programmed instruction for arc cutting does not work properly. As a result, the control may stop execution and display an alarm message. Figure 2–12 illustrates the most common mistakes other than calculation errors:

To program the arc from PT1 to PT2 in a counterclockwise direction, the correct instruction is an easy one:

G03 X3.0 Z-0.25 K-0.25;

This is enough, assuming that the previously specified feed rate is acceptable. Also, the I address is not entered, since its value is zero. However, the programmer might enter one of the following wrong instructions:

G02 X3.0 Z-0.25 K-0.25; (Wrong direction by G02. It instructs cutting toward the first quadrant, while Z and K instruct cutting in the second quadrant.)

G03 X3.0 Z0.25 K-0.25; (Wrong sign for the Z coordinate. It instructs cutting toward the first quadrant, while G03 instructs cutting toward the second quadrant.)

G03 X3.0 Z-0.25 K0.25; (Wrong sign for the K value. It assumes that the arc center is in the first quadrant, while G03 instructs the move toward the second quadrant.)

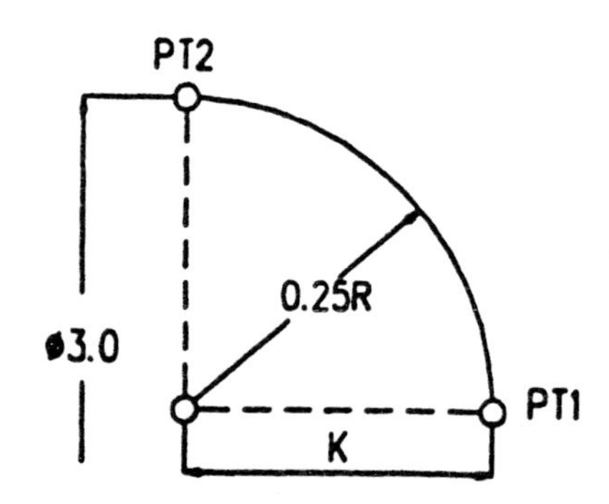

Figure 2–12 Common mistakes made when arc cutting.

In any of these cases, the control stops the execution and displays the message: "The end point cannot be viewed from the start point." or a similar one. It is a good safety feature, since the execution cannot be continued until the problem is fixed.

Dwell

The programmer enters the G04 code to program a **dwell,** which stops the tool motion at any point in the program for the amount of time specified. As the G04 code is not modal, it is in effect for the programmed block only. This function is often used when cutting a groove on the lathe to allow a full turn of the part before the tool is pulled out. It helps to achieve better part roundness and finishing. Also, after facing the part using the constant surface speed control (G96), dwell can be used to stop the tool motion until the higher RPM is changed to a lower RPM. This is sometimes necessary, because the tool moves faster from the part centerline to the new position than the spindle RPM can be adjusted. This technique is often applied when cutting hard material with a rough skin. Dwell is also used to program temporary stops in tool motion while the pull-out finger is activated for processing bar materials. On the machining center, dwell is usually programmed to stop tool feeding at the bottom of the hole to achieve a better finish. The application of programming dwell is not limited to the previous examples. There are many more occasions when dwell can improve the machining process. Note that when programming dwell, the cycle time is increased for the amount of dwell time.

Dwell is programmed together with the P, X, or U address. Which address is to be used depends on the particular control. When using the P address, the decimal point format is not acceptable.

The dwell time is expressed in thousandths of a second. Following are some examples of programming dwell:

Example 1 G04 P2000; (Dwell for 2 seconds. Decimal point format cannot be used.)

Example 2 G04 X1.5; (Dwell for 1.5 seconds.)

Example 3 G04 U3.5; (Dwell for 3.5 seconds.)

Program Control

The programmer controls the program execution using the M00, M01, M02, and M30 codes. These codes stop the program and/or rewind it to the beginning.

The programmer enters the M00 code to temporarily stop the program execution so that the operator can check the part size and condition of the tool, and to clear chips before finish cutting.

The M00 code stops the spindle rotation, coolant supply, and feeding. To continue the operations in memory mode, the programmer must reprogram the codes for spindle start and coolant supply. The spindle speed and feed rate are remembered by the control. The program execution resumes when the operator pushes the START button. The cutting tool continues operation from the same point at which it stopped.

The M00 code has to be programmed in one block by itself. The following series of instructions may be used to stop the program in order to clear chips from a bore:

```
G00 W3.0; (Pulling out the tool in the lathe Z axis in incremental.)
U3.0;(Incremental move in the X axis.)
M00; (Spindle, coolant, and feeding are stopped. Clearing chips.)
M03; (Spindle restart.)
U-3.0; (Return to position in the X axis first.)
W-3.0 M08; (Return to position in the Z axis and coolant restart.)
...; (Continuing machining.)
```

Programmers often use the M00 code to stop the program execution in order for operators to perform manual operations, such as turning the part around for an operation on the other side or moving the tailstock into position when supporting longer parts. For example:

Example 1

```
G00 X10.0 Z3.0 T0100 M09; (Turning tool returns to the start position in X and Z.
    Offset cancel by T0100. Coolant stop by M09.)
M00; (Program temporarily stopped. The operator turns the part around.)
G00 G50 X10.0 Z3.0 S1500 M42; (Coordinate system preset for the same tool.
    Speed limit of S1500 RPM. High speed range by M42.)
...; (Continuing machining.)
```

Example 2

```
G00 X8.0 Z1.0 T0800 M09; (Center drill returns to the start position. Offset cancel by
    T0800. Coolant stop by M09.)
M00; (Program temporarily stopped. The operator moves the tailstock to posi-
    tion manually or by the programmed instruction.)
G00 G50 X10.0 Z3.0 S1500 M42; (Coordinate system preset for the turning tool
    by X and Z. Speed limit of S1500 RPM. High speed range by M42.)
...; (Continuing machining.)
```

The M01 function is in fact the same as the M00 function. The only difference is that the M00 code stops the execution without any option, while the M01 code leaves the option to the operator to stop the program or not. Because of this, the M01 code is known as the optional program stop code.

There is a switch on the operator panel known as the OPTIONAL PROGRAM STOP switch. Its position determines whether or not the program execution will stop. When the switch is on, the M01 code is activated and execution is stopped. When the switch is off, the program execution continues.

After programming the optional program stop code, the programmer must reenter the codes for spindle start and coolant supply in the program. To resume operations after the optional program stop is executed, the operator must push the START button.

The M01 code is usually programmed after each tool in the program. It helps the operator when checking the program by machining the first part. If the optional stop switch is on, each tool stops after returning to the start position. The execution halts and the operator can run the next tool slowly. If he or she needs to recut the part by one particular tool, it is handy to have the same tool in a ready position. It saves time for the tool change. After the operator checks and proves the program, he or she turns off the OPTIONAL STOP switch and the execution carries out with no interruptions.

It is a good practice to program the optional stop code after the tool returns to its start position and offset is cancelled. Following is an example of how to use the M01 function effectively:

```
G00 X12.0 Z2.0; (Return the tool to its start position.)
T0300 M09; (Offset cancel for tool number 3. Coolant stop.)
M01; (Optional program stop.)
G50 X11.5 Z2.025; (Coordinate system preset for the next tool.)
```

In the preceding series of instructions, the tool is returned to its start position and offset is cancelled. This is useful when proving the program because the operator can rerun the tool immediately after the problem is solved. If the M01 code is placed before the offset cancel instruction, the operator could just reset the program without executing the block with the offset cancel instruction. When the offset is called up again, it will build up on the previous offset value because it was not cancelled. This might have serious consequences. By placing the M01 code after the offset cancel block, the programmer can avoid this possibility.

The M30 function resets all the registers to the initial state and rewinds the program to the beginning. When this code comes into effect, the spindle rotation, feeding, and coolant supply operations stop automatically. In practice, the coolant stop code M09 and spindle stop code M05 ensure faster stoppage.

The M30 code has to be programmed in one block by itself. Normally, it is the last instruction in the program, as in the following series of instructions to return the lathe tool to its start position and rewind the program:

```
G00 X15.0 Z2.0 M09; (Return the tool to its start position. Coolant stop.)
T0500 M05; (Offset cancel for tool number 5. Spindle stop.)
M1; (Optional program stop.)
M30; (Program stop and rewind.)
% (The stop code in ISO format.)
```

After the M30 code, a stop code normally appears on the monitor. It can be ER (End of Reading) in the EIA code set or % in ISO code set. When loading a program from an office computer, the M30 code causes the communication protocol to enter the stop code. When inserting the program by punching the codes directly into the memory of the machine, the programmer can see the stop code after the program registration. For instance, if he or she names a program O999, after pressing the INSERT button, he or she will see: O999%. Now the programmer can design a program between the program number and the stop code using the end of block character to separate them. The cursor usually cannot pass the stop code in a program.

The M02 function is normally used to end a program on older machines that have no computer memory but use the so-called "endless tapes." The M02 code stops the spindle and coolant as when using the M30 code, but the M05 and M09 codes are used for faster stoppage. The M02 code has to be on a line by itself, as in the following example:

```
G00 X11.575 Z1.750 M09; (The tool returns to the start position. Coolant stop.)
T0500 M05; (Tool and tool offset cancel. Spindle stop.)
M02; (Program end.)
```

This function can also be used for memory operations on newer machines, but it will not rewind the program to the beginning. Use the M99 code for that purpose.

In memory operations, it is easier to use the M30 code, but there are jobs for which it cannot be used. For example, when processing bar materials, the operator does not want the program to stop. In this case, the block delete code is programmed before the M02 code. It instructs the control to jump to the M99 instruction. Then the program is repeated until the operator turns the block delete switch off. Afterwards, the M02 reads in and the program is stopped. This is illustrated in the following series of instructions:

```
G00 T0300 M05; (Tool and tool offset cancel. Spindle stop.)
/M02; (Program end. The execution of this instruction depends on the position
    of the block delete switch. If it is on, the program does not stop; if the
    switch is off, the program stops.)
M99; (Program rewind.)
% (Stop code in ISO format, entered automatically.)
```

In memory operations, the ER in the EIA code set or % in the ISO code set must be entered after the M02 code. This is to separate the programs in the control memory.

Subprogram Control

When it is necessary to branch the execution or to repeat one operation or group of operations, a **subprogram** can be used. The subprogram is one way of performing subroutines when programming. In essence, it is another program that can be called up from the program being executed.

All newer CNC controls enable the use of subprogramming techniques. On most of the controls the subprogram manipulation is accomplished by the following codes:

M98 Subprogram call
M99 Return from subprogram

Address Description:

P Designates the subprogram number, for example 555. It must be programmed in the main program. When this address is entered in the subprogram, it designates the sequence number in the main program where the execution returns. If this address is not specified when returning, the execution is continued in the first block after the block with the subprogram call.

L Number of repeats. A subprogram is executed L times. Without the L designation, the command is executed once.

On some controls the subprogram number is designated by the address N, allowing the subprogram to be at the end of the main program. This simplifies program loading from a personal computer. On newer controls the subprogram cannot be inside or at the end of the main program. It has to be a separate program loaded into memory. The older system of subprogramming is handier because the main program and the subprogram can be seen together. But the newer system is faster since the program number has to be found in program library, not through the numerous lines of one particular program.

Following are some examples of calling the subprogram and returning to the main program:

Example 1 M98 P500 L3; (Calling the subprogram number 500, 3 times repetition.)

Example 2 M99 P123; (Returning to N123 in the main program. Note that in this case, line 123 must be numbered in order to be found. The L address is not specified; thus, the program is executed once.)

Example 3 M99; (Returning to the main program to the line after the subprogram call. For example, the part of the main program is written as: N025 G98 P101; N030 G00 X10.0 Z2.0;. Then, the return is made to N030, no matter if it is numbered or not.)

On most controls, codes other than P and L must not be programmed in the G98 or G99 blocks; but on some controls the S or F code are allowed. The program number or sequence number designated by the address P is searched from the beginning of data in memory. The execution is made from the sequence number or program number that appears first. Consequently, two identical subprogram numbers should not be programmed. The alarm sounds when the program or sequence number is not found. Up to four-digit numerals are allowed to designate a subprogram number. Leading zeroes can be omitted.

The M99 branching instruction can also be used as a GO TO command. Then the line number where the execution of the program is to be continued must be specified. The M99 instruction may be used to skip part of the data in one program searching either forward or backward.

This is the quickest way to change the order of operation sequences when machining. For example, the machining sequences are programmed as: (1) Roughing on OD (the outside diameter); (2) Finishing on OD; (3) Roughing on ID (the inside diameter); and (4) Finishing on ID. Because of the part distortion, it appears that on both OD and ID the roughing should be completed first. The program editing using the M99 instruction is the most suitable in this case:

Original Program:

N100; (Roughing on OD.)
N200; (Finishing on OD.)
N300; (Roughing on ID.)
N400; (Finishing on ID.)
M30; (Program end and rewind.)

Altered Program:

N100 (Roughing on OD.)
M99 P300; (Branching to line N300 to call the ID roughing tool.)
N200 (Finishing on OD.)
M30; (Program end and rewind.)
N300 (Roughing on ID.)
N400 (Finishing on ID.)
M99 P200; (Branching to line N200 to call the OD finishing tool.)

As per the altered program, the order of machining sequences is changed as follows: (1) Roughing on OD; (2) Roughing on ID; (3) Finishing on ID; and (4) Finishing on OD. The program execution is branched forward and backward until it ends with the M30 instruction. This change takes only a few minutes. Before running the same job again, the program should be rewritten by moving the groups of data, not by using the M99 instruction.

It is important to know how to effectively use subprogramming techniques. For example, a part is to be turned to 1.0 inch diameter and 1.1 inch long. Then 0.05-inch deep grooves by the same tool are to be cut every 0.1 inch of the length. There are 10 grooves to be cut; thus, the use of a subprogram is justified. The turning operation will be placed in the main program, while the grooving operation in the incremental mode will be placed in the subprogram.

O012; (Main program number.)
...; (Coordinate system preset, tool and tool offset call, and spindle start are omitted.)
G00 X1.0 Z0.05 M08; (The tool is positioned to cut a 1.0 inch diameter. Coolant on.)
G01 Z-1.1 F0.01; (Turning 1.0 inch diameter, 1.1 inch in length.)
X1.05; (Moving the tool 0.05 inch away from the part on the X axis.)
G0 Z0; (The tool is moved to position from which the subprogram is called up.)
M98 P333 L10; (Calling the subprogram 333, 10 times repetition.)

```
...; (The tool is returned to its start position and offset is cancelled.)
M30; (Program end.)

O333; (Subprogram number.)
G00 W-0.1; (Positioning in the Z axis to cut the first groove.)
G01 X0.9 F0.008; (Cutting the groove in the X axis. The X coordinate is calcu-
    lated as follows: 0.05 • 2 = 0.1. This is subtracted from a 1.0 inch
    diameter.)
G00 X1.05; (Return to the start position on the X axis.)
M99; (Return to the main program.)
```

The distance between the grooves is kept by programming W-0.1 in the incremental mode in the subprogram. This increments the tool position each time the subprogram is called up. The X value (X0.9), specified as the groove diameter, is the same for all of the grooves. Thus, it can be programmed in the absolute mode. Before calling the subprogram, the tool is positioned at Z0. This is the case if the first groove is a distance of 0.1 inch from the part face. If this distance is, say, 0.08 inch, the tool start point in the main program would be Z0.02. This is because 0.08 + 0.02 = 0.1, which would be the value programmed by the W. This would make sure that the first groove is machined 0.08 inch from the part face, but the distance between the rest of the grooves is 0.1 inch.

This was a rather simple example, but the advantages of using subprogramming techniques are obvious:

1. The program is shorter, resulting in fewer possibilities for programming errors.
2. The program is easier to edit.
3. The program is easier to read by the operator.
4. The program needs less memory for loading.

The main program can branch to several subprograms as many times as desired. This means that there is no limit for the number of subprogram calls and returns. It is limited only by the machine memory capacity. The following illustrates calling three subprograms from the main program:

```
O010; (Main program number.)
M98 P300; (Calling the subprogram O300.)
M98 P301; (Calling the subprogram O301.)
M98 P302 L2; (Calling the subprogram O302, 2 times repetition.)
M30; (Program end.)
```

This could be a long and complex program even though it is written in a few lines. The real action is placed in the subprograms numbered O300, O301, and O302. In order to be executed, they must loaded into the memory of the CNC machine. The execution starts with subprogram O300. When it is finished, the control returns to the main program. Then, the control branches to subprogram O301. After returning, subprogram O302 is called up and repeated twice. Afterwards, program execution returns to the main program, which ends with the M30 code.

The advantage of this technique is that data often repeated may be placed into the subprogram and used when needed. Another advantage is that the subprograms may be used for several main programs, for different jobs. Then they can be called up in a different order with a different number of repetitions.

A subprogram call from another subprogram is known as **nesting.** On some machines it is ignored, but some newer machines allow nesting. A subprogram may be nested several levels, depending on the capabilities of the control. Nesting is an advanced programming technique, but there is nothing mystical in it. The program execution goes to the lower levels when branching. When returning, it goes to the upper levels and stops at the main program where it all started. For example:

```
0005; (Main program number.)
M98 P100; (Calling the subprogram O100.)
M30; (Program end.)

O100; (Subprogram number.)
M98 P500; (Calling the subprogram O500, first level of nesting.)
M99; (Return to the main program.)

O500; (Subprogram number.)
M99; (Return to O100.)
```

The subprogram O100 could be used to cut several grooves, and the subprogram O500 could be used to break the edges on the grooves.

Programming Efficiently

Say a program is ready and stored into the memory of the CNC machine. Suppose that all the codes are correct, as well as the tool path, according to the part geometry. But is the program good, or even the best possible? Will it satisfy the main requirement in programming CNC machines?

The accuracy is built into the machine, and the programmer just uses it. So what is the main requirement in programming CNC machines? Consider the two positions in Figure 2–13.

The program is supposed to drill six holes on a bolt circle. At position (a), the tool starts drilling hole 1 in the first quadrant; then it goes to drill hole 5 in the second quadrant. The coordinates are the same, and only the signs are different. The program continues in a zigzag direction because the programmer may feel this way of repeating the coordinates and changing the sign is easier.

At (b), the tool is programmed to drill the holes in a circular motion, from hole 1 to hole 6. Which program is faster and more efficient? Obviously, the program for position (b). More time is wasted when drilling in a zigzag motion than when drilling in a circular motion. It is more noticeable when machining the holes on a

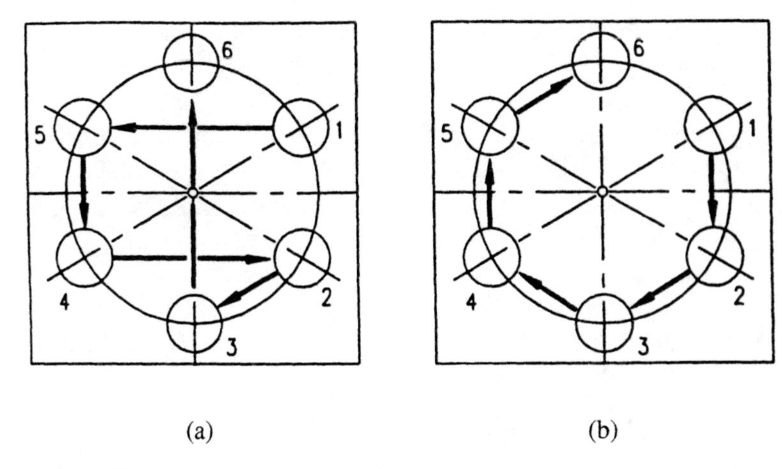

Figure 2–13 Efficiency when programming.

large bolt circle. The program for position (b) is also safer for the operator since there are no unexpected back and forth motions. When the work orders range in the hundreds or thousands, a considerable amount of time and money would be lost if the program is designed like position (a).

The best possible program does the job in the shortest time and the safest way. How fast and safe the program is is a measure of its efficiency. For one technology operation the program can be written many ways, but only one or two will make it possible to achieve the full efficiency of the machine. That is the main requirement when programming CNC equipment.

Basic Program Structure

Now you have a clear concept of how to establish the reference points and how to use the different programming functions. You know the syntax of CNC language, as well as the main requirement in programming CNC machines. You are ready to design your first program.

Note that the part program for each tool has the same basic structure. This means that the same programming steps have to be taken when programming, regardless of the tool being used. The following lists these steps in a consecutive order:

1. Establish the coordinate system.
2. Call the tool and tool offset.
3. Turn the spindle.
4. Approach in rapid mode.
5. Approach by faster feed rate.
6. Program the cutting mode.
7. Return the tool to its start position and cancel the offset.

The first program in this text is based on creating the simple part illustrated in Figure 2–14. The material is Mild Steel. It is assumed that the G20 and G99 codes are set by the parameter, so inch data input and feed per revolution will be generated. It is also assumed that the parameter is set so that programming on diameter can be used.

O30; (Program number.)
N10 G50 X10.0 Z2.0 S1300 M42; (Coordinate system preset. Speed limit of 1300 RPM. High speed range by M42 code.)
N20 G00 T0100; (Indexing the tool 01 in a ready position.)
N30 G96 S550 M03; (Spindle start CW by M03. The G96 code instructs the control to use the constant surface speed of 550 inch/min. At this moment, the spindle is turning slowly since the X position is 10.0 inch.)
N40 G00 X2.0 Z0.1 T0101; (Approach in rapid mode by the G00 code. The tool is brought into position to cut a 2.0 inch diameter in the X axis and 0.1 inch from the face in the Z axis. Tool offset is called by the T0101 instruction. Notice that the last two digits are changed. They were 00 when the tool was called in the N20 block. At this point, the spindle is turning fast according to a small diameter of 2.0 inch.)
N50 G01 Z0 F0.015 M08; (Approach in the Z axis by a slightly faster feed, not by rapid. Coolant on by the M08 code.)
N60 Z-1.0 F0.011; (Cutting a 2.0 inch diameter up to the shoulder by the appropriate feed rate.)
N70 X3.1; (Cutting a 3.0-inch diameter shoulder; it is programmed for a bit larger diameter to allow the tool to leave the part in the same block.)
N80 G00 X10.0 Z2.0 T0100 M09; (The tool is returned to the start point by programming the same coordinate as in the G50 block. The tool offset is

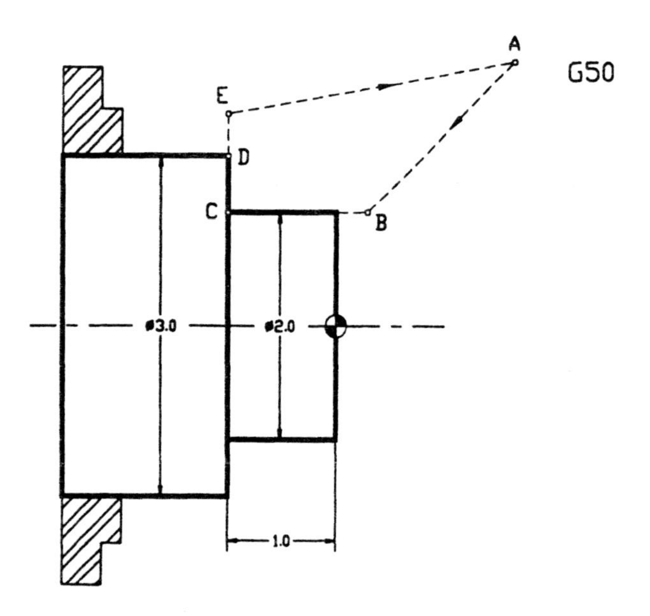

Figure 2–14 Part used to illustrate a basic program structure.

cancelled by entering 00 instead of 01 for the last two digits. Coolant stopped by the M09 code.)
N90 M30; (Program stop and rewind.)

As stated earlier, each block has to end with (;), the character for the end of block code when using ISO format. Note that the end of block code must show up on the monitor, but it does not appear on the program hard copy. The program must not contain any code that is not part of the code set used.

When creating the program for one machining operation, always plan in advance the next operation on the same part, if any. The same precautions for chucking, clamping, finishing allowance, and so on should be exercised in the same way as for manual machining. The program is unusable if, after cutting the thread on the first side, the part cannot be chucked to start machining on the other side. (This is a rather simple example in order to stress the importance of planning when programming.)

To keep the part in size, the order of machining sequences may be very important. This is especially true when programming machining sequences for complicated parts, because the part origin is established more than once. Before programming, always carefully study the blueprint to find the best order of machining sequences. Each part is specific. If the part is heavy, it might be a good idea to complete the roughing first in order to remove the excess material. It will make it easier for the operator to handle the part. If the part is complicated, it is better to make several short programs for several operations. Trying to write one long program may result in confusion for both the programmer and the operator. Long programs are not easy to handle, thus they are not safe either.

Program Loading

Once the program is written it is loaded into the memory of the CNC machine. There are two ways of loading the program:

1. Direct data input
2. Indirect data input

A method of direct data input is **manual data input (MDI).** It can be achieved via the CNC machine keyboard, push buttons, and switches. Typing the program into the memory via the machine keyboard can be a very fast and efficient process for loading short programs. An experienced programmer/operator can get these programs ready to run faster than using other methods. This method is especially efficient if the machine has multitasking capabilities which allow execution of a program as a foreground operation and creation of a new program as a background operation.

When loading long programs, indirect data input is more efficient. It is achieved using communication channels, floppy disks, punched tapes, magnetic tapes, and punched cards. Communication channels and floppy disks are used in conjunction with personal computers that are linked to the machine. The computer link also allows **direct numerical control (DNC)** to be used. In direct numerical control, a central computer, known as a host computer, controls the operation of one or more machines. Punched tapes, magnetic tapes, or punched cards can be produced using a teletypewriter or punch unit. They are read by the machine's electromechanical or photo-electric tape readers, but this is an old technology not used much anymore. Punching the information directly onto the machine keyboard and the more indirect PC link are the two methods of input widely accepted these days.

Program Proving

It is always possible to make mistakes when programming, either in coding or calculating. There are two steps for checking the program. First, the program has to be checked for errors in coding, positioning, and so on. Secondly, the program performance has to be proven on the machine. One way of checking the program when incremental programming is used is to calculate the sum of tool movements on each axis in order to return the tool to its start position. If the sum is zero, it means that the tool is returned properly. This can be done by manual calculations or by part programming software, if available.

When checking the program on the machine, the operator uses the **machine Lock function.** It turns on the spindle, but the tools do not move. The program runs on the monitor and every format error results in an alarm message. This way of checking the program is widely used in demonstrations to students.

A better way of checking the program is by "cutting the air." This may be done by removing the part from the machine, or the tool may be moved away from the part. Also, the machine **Dry run function** may be used: The tools move by the specified feed rate set by the parameter, even when rapid traverse is programmed.

When machining the first part, the feed rate may be slowed down when the tool is cutting close to the chuck or fixture. In this case, the feed rate override switch may be used. After comparing how far the tool is from the chuck or fixture (either by sight or by measuring it after stopping the spindle), and Distance to Go value on the screen, the execution should be stopped if anything seems to be wrong. Then, the program and setup must be checked for any discrepancy.

It is strongly recommended that you use the machine dry run to prove a new program. Often the proven program is also dry run for a new setup, especially if the program was not used for a long time. Program execution by a single block is also recommended. It allows the operator to read and confirm each line in the program before its execution takes place.

There are some important practical points which should be observed by the operator when checking the program before executing it:

1. Are the same numbers programmed in the coordinate system preset block and in the block that returns the tool to the start position?
2. Are the numbers used to call and cancel the tool and tool offset the same? For example, in the program the end mill may be labeled as tool number 2. On the machine, this tool is already set at position number 5. It is easier to change the tool number than the tool position. When you change the tool and tool offset number in the block where they are called up, do not forget to do the same in the block where they are cancelled.
3. Is the command to return the tool to its start position given after the tool is pulled out from the hole or bore, or it is programmed immediately after the cutting is finished, while the tool is still inside the part? The tool must be programmed to pull out before it is returned to the start position.
4. In the lathe program, there must not be any X coordinates with a minus sign. The X coordinate has a small negative value only when facing a bar stock, because of the tool nose radius. This is opposite when programming for the lathe with the left-hand coordinate system.
5. Is there a G00 code that will generate an unexpected rapid motion instead of feeding as it's supposed to? This is particularly important when the tool is close to the part, fixture, or chuck.
6. What is the longest motion on the lathe Z axis? Compare it with the length of the part stickout to prevent the tool feeding into the chuck. If needed, pull the part out more and establish a new part origin.

If the programmer forgets to enter the codes for activating the coolant or starting the spindle, it can be easily noticed and fixed. If the feed rate is excessive and the surface finishing is not satisfactory, it can be changed for the next part. But any of the situations listed above can result in serious problems. Make every effort to develop the ability to quickly detect dangerous bugs in the program.

Summary

Absolute and incremental positioning are used in programming CNC machines. In absolute programming, all measurements are made from the part origin and any coordinate programmed has an absolute value in respect to the absolute coordinate system zero point. The machine control uses the part origin as the reference point in order to position the tool during program execution. The sign of each coordinate depends on where the tool is moving according to quadrants. In incremental programming, the tool movement is measured from the last tool position. The relative distance between the current location and the next programmed point is entered into the program. The sign of each coordinate depends on where the tool is moving with respect to the machine axis.

The machine parameters are selected instructions by which the machine is programmed, set up, and operated. There are many machine functions, such as the feed rate mode, type of programming, the amount of rapid traverse rate for each axis, whether the dry run is effective or not, and the data input mode selection, that can be affected by the parameter setting. Older machines have fewer parameters that can be set or changed by the programmer or operator. Newer machines are much more sophisticated and have hundreds of parameters. In general, the more parameters the machine has, the more flexible and sophisticated it is.

Programming functions are the machine codes G and M, and some switches. For instance, each tool in the program must have its coordinate system preset block in order to establish the program origin. For that purpose, the G50 function on the lathe and G92 function on the machining center are usually used. When reading the coordinate system preset block, the control assumes that the values entered describe the tool distance from the part origin in respect to each machine axis. On the lathe, the G50 function is also used to set the maximum spindle RPM, which will not be exceeded even if a higher speed is entered later in the program.

The T function is used to call the particular tool and tool offset in the program. Tool offset is the function used to correct the values entered in the coordinate system preset block. This can be done quickly on the machine without actually entering the program and changing the values. Using the tool offsets, it is easy to set up the tools and to make adjustments in the part size.

There are two spindle speed modes used on CNC machines, revolutions per minute (RPM) and constant surface speed. The spindle speed in revolutions per minute is also known as direct RPM or constant RPM. The change in the tool position does not affect the RPM commanded. This means that the spindle RPM will remain constant until another RPM is programmed. On the majority of machining centers, constant RPM is the only spindle speed mode used. On most of the lathes, the constant RPM is programmed using the G97 code. Constant surface speed is almost exclusively used on lathes, and it is programmed using the G96 code. In this mode, the RPM changes according to the diameter being cut: The smaller the diameter, the more RPM is achieved; the bigger the diameter, the less RPM is commanded. This is changed automatically by the machine speed control unit while the tool is changing its positions. This is the why spindle speed mode is also known as the diameter speed.

Cutting operations may be programmed using two basic feed rate modes, feed rate per spindle revolution and feed rate per time. The feed rate per spindle revolution depends on the RPM programmed. This type of the feed rate is almost exclusively used on lathes. The feed rate per time depends on the feed rate specified. The tool will advance for the amount of the feed rate programmed, regardless of the spindle revolution. This type of feed rate is usually used on machining centers and it is expressed in inch/min. in inch data input, or m/min. in metric data input.

The machine tool can be positioned in rapid mode using the G00 function or in cutting mode using the G01, G02, or G03 codes. Each of these codes is modal—once programmed they stay in effect until cancelled. The G00 function is normally used for positioning when the tool is approaching or leaving the part.

It is also used to call or cancel the tool and tool offset, as well as to cancel a canned cycle.

The G01 function is called straight cutting, linear cutting, or linear interpolation. In this cutting mode, the tool is moved from point to point by the programmed feed rate. Using this function, we normally cut a straight line, taper, chamfer, or groove, but when needed, a radius, arc and special curves can be cut too. Also, a cut-off can be performed. The G02/G03 function is used for circular interpolation to cut an arc, arc segment, or even a full circle. The tool is moved by the two axes simultaneously. The arc can be cut in a clockwise (G02) or counterclockwise (G03) direction. After placing the tool at the arc start point, the following three pieces of information are required to cut an arc: cutting direction, G02 or G03; arc end point coordinates; and arc center displacement values (I, J, and K values).

Control of the program execution is possible when using the M00, M01, M02, and M30 codes. These codes are used to stop the program and/or to rewind it to the beginning.

A subprogram can be used when it is necessary to branch the execution or to repeat one operation or group of operations. In essence, it is another program that can be called up from the program being executed. All newer CNC controls enable the use of subprogramming techniques. On most of the controls the subprogram manipulation is accomplished using the M98 code for subprogram call and the M99 code for return from subprogram.

Once the program is written, it is loaded into the memory of the CNC machine. There are two ways of loading the program, direct data input and indirect data input. One method of direct data input is known as manual data input (MDI). It can be achieved via the CNC machine keyboard, push buttons, and switches. The indirect data input is achieved using personal computers linked to the CNC machines, although some older systems still use tapes.

It is always possible to make mistakes when programming, either in coding or calculating. Therefore the program has to be checked for errors in coding and positioning, and the program performance has to be proven on the machine. When checking the program on the machine, the machine Lock function may used. It turns on the spindle, but the tools do not move. The program runs on the monitor and every format error results in an alarm message. A better way of checking the program is by "cutting the air." This may be done by removing the part from the machine, or the tool may be moved away from the part. The Dry run function may also be used. When the program is dry run, the tools move by the specified feed rate set by the parameter, even when program execution should be stopped if anything seems to be wrong. Then, the program and setup must be checked for any discrepancy.

It is strongly recommended that the the machine Dry run be used to prove a new program. Often the proven program is also a dry run for a new setup, especially if the program was not used for a long time. Program execution by a single block is also recommended. It allows the operator to read and confirm each line in the program before its execution takes place.

The basic program structure for any cutting tool is the same. After the program is created for one tool, the same principle can be applied for the other tools in the program. Normally, the specifics of each cutting tool must be taken into consideration.

There are certain rules to be followed when using CNC syntax. Even though these rules are much the same for all of the CNC machine tools, different machine builders might have slightly different usage of codes and formats. Thus, when programming CNC equipment, it is wise to check the machine manual whenever in doubt. Also, when using any programming function, the programmer must have a clear understanding of what it does when executed.

Key Terms

absolute programming
arc center displacement
circular interpolation
constant RPM
constant surface speed
coordinate system preset
direct numerical control (DNC)
Dry run
dwell
feed rate per spindle revolution
feed rate per time
inch data input
incremental programming
linear interpolation
machine Lock function
metric data input
multiquadrant circular interpolation
nesting
parameters
programming on diameter
programming on radius
slash code
spindle speed limit
subprogram
tool offset

Self-Test

Answers are in Appendix E.

1. In ________________ all measurements are made from the part origin.
2. In ________________ the tool movement is measured from the last tool position.
3. ________________ are selected instructions called defaults or first choices by which the machine is programmed, set up, and operated.
4. The ________________, also known as block delete or optional skip code, is a very helpful function that may be used to solve the most difficult programming tasks.
5. Each tool must have its ________________ block in order to establish the program origin.

6. The ______________ is the maximum spindle RPM which will not be exceeded even if a higher speed is entered later in the program.
7. ______________ is the function used to correct the values entered in the coordinate system preset block.
8. The ______________ is the spindle speed in revolutions per minute.
9. When using the ______________, the spindle RPM changes according to the diameter being cut.
10. ______________ depends on the RPM programmed.
11. ______________ depends on the feed rate specified.
12. ______________ is used to program straight cutting between two points.
13. When using ______________, the tool is moved by the two axes in simultaneous motion.
14. The ______________ is the incremental distance from the arc start point to the arc center.
15. Cutting an arc placed in more than one quadrant is known as ______________.
16. The ______________ is programmed using the G04 code to stop the tool motion.
17. A ______________ is another program that can be called from the program being executed.
18. ______________ is a subprogram call from another subprogram.
19. In ______________, a central computer controls the operations of one or more machines.
20. When using the ______________, the program runs on the monitor and the spindle turns on, but the tools do not move.
21. When using ______________, the tools move by the specified feed rate set by the parameter, even when rapid traverse is programmed.

Relating the Concepts

No answers are suggested.

1. Which programming modes are used in programming CNC machines?
2. Differentiate between incremental and absolute programming.
3. List some of machine functions that can be affected by the machine parameters.
4. Explain the use of the coordinate system preset block.
5. Define the tool offset.
6. Differentiate between constant RPM and constant surface speed.
7. Differentiate between the feed rate per time and per spindle revolution.
8. List the cutting modes in CNC programming.
9. Which three pieces of information are necessary when programming an arc?

10. Which codes are used to control the program execution?
11. Describe how the M01 code works.
12. Identify the use of the M98 and M99 codes.
13. Explain how a program can be loaded into the CNC machine.
14. Describe the use of the machine Lock function.
15. The program can be proven by "cutting the air." Explain the meaning of this phrase.
16. Describe the Dry run function.
17. List the programming steps to be taken when programming, regardless of the tool being used.

3 Machine Setup

Key Concepts

Home Position
- Home Position on the Lathe
- Home Position on the Machining Center

Coordinate System Preset
- Coordinate System Preset for Shaft Work
- Work Coordinates

Tool Offset Consideration
- Tool Length Offsets
- Geometry Offsets
- Calling and Cancelling Tool Offset
- Tool Offset Adjustment

Methods of Programming the Coordinate System
- Real Values
- Same Imaginary Values
- Approximate Real Values

Setting Up the Tools on the Lathe
- The Imaginary Tool Tip Method
- The Tool Nose Center Method

Setting Up the Tools on the Machining Center

Setup Information

Screen Reading
- Offset Screen
- Work Zero Offset Screen
- Position Screen

Home Position

After the program is designed and loaded into memory, the machine has to be set up with the tools set according to values specified in the program. The point from which the tools start program execution and to which they return is known as the home position. This is where the tool changes usually take place, so the home position is also known as the *tool change station.* On some machines, especially mills, the tool change position is dictated by the machine design.

The **home position** is established according to the machine origin, which is set by the machine tool builder and normally cannot be changed. The home position may be set at any convenient point inside the machine's electronic and mechanical limits, even at the machine origin. All of the tools have to be clear from the part when the tool changes occur.

When the home position coincides with the machine origin, it is easy to position the tool at its start point, either when the machine is turned on or after program execution is interrupted. The machine **Zero Return function** may be used for this purpose.

When the home position does not coincide with the machine origin, the operator needs to know the location of the home position. Then he or she can manually perform the return to the home position if he or she has not changed the zero reading since it was first set. In actual practice, the operator should not depend on the screen reading since it may be changed during machining. He or she must be certain that any time when needed, the machine may be placed at the home position. The following procedure may be used to establish the home position.

1. Place the machine at the location chosen as home position, and set the zero reading.
2. Return the machine to the machine origin by selecting each axis by the AXIS SELECT switch and using the ZERO RETURN button.
3. Remember or write down the distance.

When in doubt if the home position is correct, the operator should zero the slides and move the machine a known distance.

Which should be established first, the home position or the part origin? Although the operator can choose to set either first, it is more efficient to establish the part origin first. Then the operator can set the home position at any point close to the part.

Home Position on the Lathe

On small CNC lathes, the home position usually coincides with the machine origin. On medium-sized lathes, the machine origin is used as the home position on the X axis, while on the Z axis the home position is set a distance away from the part face. When machining short parts on longer lathes, the home position should be set between the machine origin and the part origin. It cuts down the time it takes for the tool to travel from the machine origin toward the part.

For most jobs on the lathe, the operator keeps the same home position on the X axis, while for the Z axis, he or she changes the home position according to change in the part origin. Figure 3–1 illustrates the relationship between the home position and the machine origin on the lathe.

Figure 3–1 shows that the home position is related to the machine origin by the U and W coordinates. By knowing these coordinates, the operator is able to position the machine at the same location chosen as the home position. On the monitor, the distance from the machine origin to any new position is expressed as the machine relative distance. The home position is also related to the part origin by the X and Z coordinates. On the monitor, the distance from the part origin to any new position is expressed as the machine absolute distance.

The operator must be told which tool to use to establish the home position. To set the home position in such a way that all of the tools are clear from the part and the machine, he or she usually uses the longest tool. This tool is known as the **master tool.** Any tool may be used as the master tool, but for safety reasons, the longest tool is most often used. Then, if this tool is clear, all of the other tools will be clear as well. To establish the home position on the lathe:

1. Face the part by turning the tool to establish Z0, the part origin point on the Z axis.
2. Bring the longest tool and touch off the part face.
3. Set the zero reading on the screen: On older machines, enter Z0; on newer machines, enter W ORIGIN.
4. Move the tool away, say, for 2.0 inches. This distance is usually the Z value programmed in the coordinate system preset block for the master tool. Set the zero reading again. This is the home position on the Z axis. Move the machine

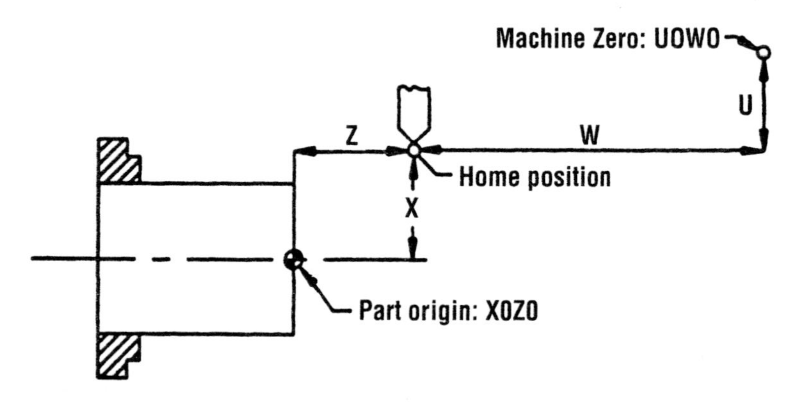

Figure 3–1 The home position on the lathe.

to its origin on the Z axis. When the light comes on, take a reading. This is the distance from the home position to the machinc origin on the Z axis.

5. Move the machine to its origin on the X axis. When the light comes on, enter X0 or U ORIGIN. This is the home position on the X axis, which coincides with the machine origin.

To set the home position closer to the part on the X axis, the operator would change the procedure in step 5 as follows:

- Move the tool away from the part to clear on the X axis.
- Set the zero reading.
- Move the machine to the origin on the X axis.
- Take a reading. This is the distance from the home position to the machine origin on the X axis. In this case, the home position does not coincide with the machine origin.

The other tools should be set up after the home position is established by the master tool. The operator then sets up the tools with respect to the master tool. Should the home position be changed for a new part to be machined, only the master tool's distance from the machine zero needs to be changed.

The programmer can reference the home position on the lathe automatically in one of two ways: He or she can design a small program on the end of the current program, or he or she may load a separate program into the memory. Then, the operator can perform this task using the G28 function, as below:

```
...; (Machining.)
M30; (Program end and rewind.)
N999; (Block number.)
G28 U0; (Return to the machine origin on the X axis first for safety.)
G28 W0; (Return to the machine origin on the Z axis.)
G00 W-12.123; (Move distance from the home position to the machine origin on
    the Z axis.)
M30; (Program stop and rewind.)
```

The program stops and rewinds after the first M30 code. If the operator wants to return the machine to the home position, he or she calls up the N999 block. Then, the machine zeroes slides and returns to home position automatically. When the second M30 code reads in, the program stops and rewinds to the beginning. The machine is ready to machine the next part. When programming the G28 function, the programmer must enter only the relative coordinates U and W. Both have a minus sign. If the home position does not coincide with the machine origin on the X axis, the programmer enters the address U as the move distance from the home position to the machine origin on the X axis. For example:

```
G00 U-2.0 W-12.123;
```

On a machine with a left-hand coordinate system, the programmer enters the U address with a plus sign, as illustrated in the following series of instructions:

01234; (Program number for the separate program loaded into memory.)
G28 U0 W0; (Return to machine origin on both axes simultaneously.)
G00 U1.5 W-11.785; (Move distance from the home position to the machine origin on X and Z axes.)
M30; (Program stop and rewind.)

On the X axis, the operator usually keeps the home position on the machine origin, while on the Z axis, he or she moves the home position according to the part overall length. To start machining on a shorter part, the operator relocates the home position farther from the machine origin by entering a larger minus W value. To start machining on a longer part, he or she moves the home position closer to the machine origin by entering a smaller minus W value. This is true when using the machine with either coordinate system, left-hand or right-hand.

On some newer lathes, the operator uses the machine Shift function to establish the home position. This function allows shifting the home position to a new location from the Shift page on the screen. Then, the operator does not need to search for and to call the program with the G28 function to return the machine to the home position.

Home Position on the Machining Center

On the machining center, there are no special addresses to express the distance from home position to the machine origin. The programmer uses the same X, Y, and Z addresses to establish both the home position and the part origin. X and Y reference the machine table, while Z references the spindle.

On the Z axis, the operator usually keeps the home position at the machine origin. However, when looking to improve cycle time, he or she may set the home position closer to the part. For the X and Y axes, most of the time the operator sets the home position at the same location as the part origin.

To establish the home position on the machining center, the operator usually sets the part origin first. He or she loads an edge finder or dial indicator into the spindle, and in manual mode brings the spindle to the location that the programmer has indicated as the part origin. Afterwards, the operator zeroes the slides on the X and Y axes and takes the reading. He or she writes down the values on the screen and makes the return to the home position manually when needed. To establish the home position on the machining center:

1. Load the dial indicator or edge finder into the spindle.
2. Touch off the part and find the part center or any other location chosen as the part origin. Use the ORIGIN button or manually enter X0,Y0.
3. Move the tool to the machine origin on the X and Y axes using the ZERO RETURN button.
4. When the lights come on, take a reading. This is the distance from the home position to the machine origin on the X and Y axes.

5. Move the tool to the machine origin on the Z axis using the ZERO RETURN button.
6. When the light comes on, use the ORIGIN button or manually enter Z0. This is the home position on the Z axis, which coincides with the machine origin.

As true for the lathe, the programmer can implement the automatic referencing of home position on the machining center. He or she uses the G28 function, as shown below:

```
G28 Z0; (Return to the machine origin on Z.)
G28 X0 Y0; (Return to the machine origin on X and Y.)
G00 G91 X-20.5 Y-9.5; (Rapid traverse to the home position on X and Y axes in
    incremental.)
M30; (Program end and rewind.)
```

For safety's sake, the programmer should program the return to the machine origin on the Z axis first, then on the X and Y axes. The tool might hit the part or fixture if he or she programs the return on all three axes at the same time.

Coordinate System Preset

Each tool in the program must have the coordinate system preset block. It is known as **presetting the register.** For that purpose the programmers most often use the G50 function on the lathe and the G92 function on the machining center, although there are other ways of presetting the register.

Until the coordinate system is preset, the control has no information about the location of the tool in reference to the part origin. Each tool may have a unique preset value. In order to enable the machine to use the tool length offset properly, the programmer must enter the coordinate system preset block before the block with the tool offset call.

When executed, the coordinate system preset block tells the control how far the tool is from the part origin according to each machine axis. If tools were of the same length, the coordinate system preset block would contain the same values for all tools. In reality, though, the tools always differ in length, even by small amounts. Consequently, the longer tool is closer and the shorter tool is farther from the part face where the part origin may be set (Figure 3–2).

According to Figure 3–2, the programmer may enter the following coordinate system preset blocks for the tools shown:

```
Tool 1    G50 X16.0 Z8.0
Tool 2    G50 X13.0 Z3.0
```

This means that the tip of Tool 1 is 8.0 inches away from the part origin on the Z axis and 16.0 inches (diameter value) from the part origin on the X axis. The tip of Tool 2 is 3.0 inches away from the part origin on the Z axis and 13.0 inches from the

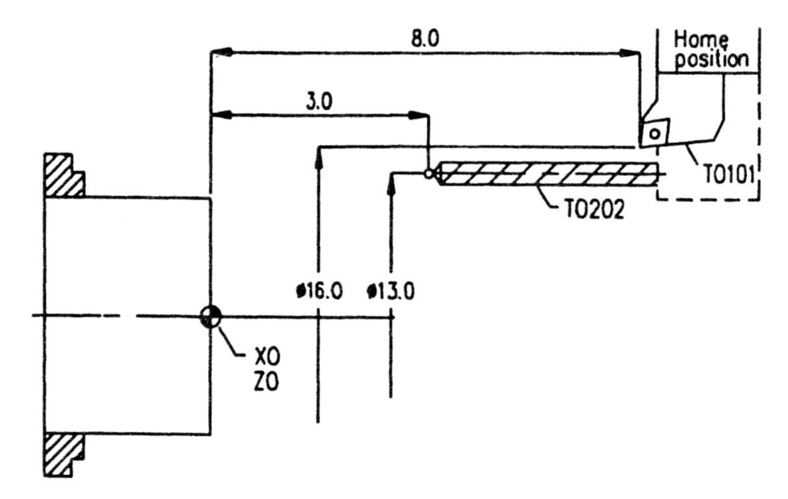

Figure 3–2 The coordinate system preset on the lathe.

part origin on the X axis. When the control reads the block with the coordinate system preset instruction, there is no tool movement; it simply accepts the values entered as the distances from the tool tip to the part origin.

When programming the G50 for the lathe or the G92 for the machining center, the programmer should take the following precautions into account:

1. The values programmed must enable all tools to start from and return to the same tool change station.
2. The values for the G50 or G92 block must be programmed as absolute, not incremental. Consequently, U and W addresses may not be used on the lathe.
3. If the wrong values are entered for the G50 or G92 block, the tool may not reach the programmed position or it might travel too far, causing a crash. The same might occur if the tool does not start the cutting cycle from the home position. The control assumes that the tool is at the home position, even if it is not.
4. The RPM limit set by the G50 function is effective only in the G96 mode. The G97 mode has no effect since it uses the constant RPM programmed. Thus, there is no need for the speed limit, but the programmer must take care that the RPM does not exceed that set by the machine tool builder.

Coordinate System Preset for Shaft Work

When machining the shafts or any longer parts on the lathe, the tailstock center is used for support. The programmer should make sure that the tools do not interfere with the tailstock when approaching and leaving the part. There are two widely accepted ways of programming the coordinate system preset for shaft work. One way is to program a small value, or even a zero value, for the Z axis when presetting the coordinate system (Figure 3–3).

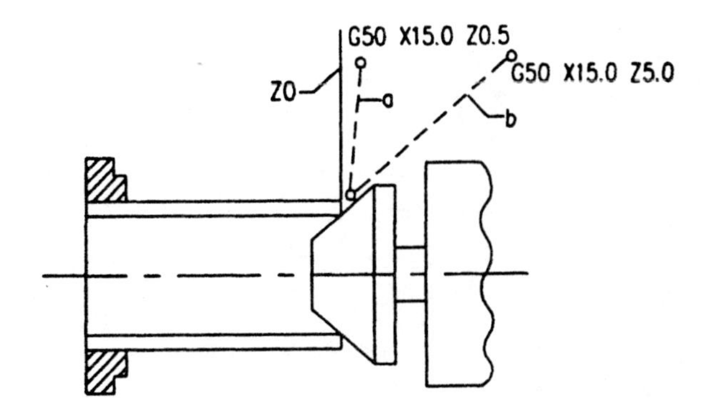

Figure 3–3 Presetting the coordinate system for shaft work.

When the tool travels along line a, there is no interference between the tool and tailstock because the Z value in the coordinate system preset block is only 0.5 inches. It enables the tool to move down in an almost perpendicular line. If the programmed Z value was 5.0, the tool would move along line b. It is likely that the tool would then collide with the tailstock.

Another way of presetting the coordinate system for shaft work is to avoid simultaneous movement of the slides when the tool is approaching or leaving the part. Thus, when approaching the part, the first move of the tool would be programmed in the Z direction, then in the X direction. When the tool returns to the start position, these commands are reversed: The tool retracts first on the X, then on the Z axis, as illustrated in the following series of instructions:

```
G50 X14.0 Z5.0 S1500 M42; (Coordinate system preset. Speed limit. High range.)
G00 T0101; (Tool and tool offset call.)
G96 S575 M03; (Spindle start.)
G00 Z0.2; (Positioning on the Z axis to 0.2 inch from the part face.)
X4.0; (Positioning on the X axis to 4.0 inch diameter.)
...; (Machining.)
G00 X14.0; (Return on the X axis first.)
Z5.0; (Return on the Z axis.)
T0100 M05; (Offset cancel. Spindle stop.)
```

This technique slows down cycle time a bit since the tool does not approach and leave the part in one diagonal move, but it also proves to be very safe.

Work Coordinates

When programming a number of identical parts, the programmer either establishes the part origin once or a number of times. To establish the part origin once, he or she usually creates a subprogram using incremental programming. It is more complicated to establish a greater number of part origins during machining. That is why machine tool builders have introduced the G54, G55, G56, G57, G58, and G59 codes known as **work coordinates** or **fixture offsets.** The work coordinates allow assignment of multiple part origins in one program.

On the machining centers, the programmer may use the work coordinates to establish a new part origin when repeating the same operations at different locations, or to program different operations at different locations. The work coordinates are also handy in lathe programming, when machining one part before cut-off and continuing on a new part. One popular control allows selection of six standard work coordinate systems:

G54 Work coordinate system 1 selection
G55 Work coordinate system 2 selection
G56 Work coordinate system 3 selection
G57 Work coordinate system 4 selection
G58 Work coordinate system 5 selection
G59 Work coordinate system 6 selection

The work coordinates are also useful in small runs when several jobs should be finished in a short period of time. Then, the operator assigns a different work coordinate system for each new setup. Afterwards, switching back and forth from one program or job to another is easy, just by calling the program into memory. The control remembers the assigned work coordinates for each setup.

Following is the programming concept using the work coordinates when several part origins need to be established in one program to perform the same job on several locations.

```
G00 G54 X0 Y0; (First coordinate system preset by the G54. Positioning on the
    X and Y axes.)
M98 P333; (Calling the subprogram to perform a particular task.)
G00 G55 X10.5 Y0; (Second coordinate system preset by the G55. Positioning
    on the X and Y axes.)
M98 P333; (Calling the subprogram to perform a particular task.)
G00 G56 X15.5 Y3.0; (Third coordinate system preset by the G56. Positioning
    on the X and Y axes.)
M98 P333; (Calling the subprogram to perform a particular task.)
M30; (Program end and rewind.)
```

The values for the G54, G55, and G56 functions are not entered in the program. The operator uses manual data input (MDI) to set these values, which are referred to the machine origin. For example:

G54 X-8.0050	Y-5.8750	Z0.0000
G55 X-16.1200	Y-5.8750	Z0.0000
G56 X-24.3260	Y-4.8750	Z0.0000

On most machines the work coordinates are entered with a minus sign since the tool can only move in a minus direction from the machine origin. When executing the program, the control reads these values and establishes each work coordinate system.

To find these values, the operator sets a zero reading at a position where a particular part origin will be set, and then zeroes the slides. The screen reading on a particular axis contains the work coordinates for that axis (Figure 3–4).

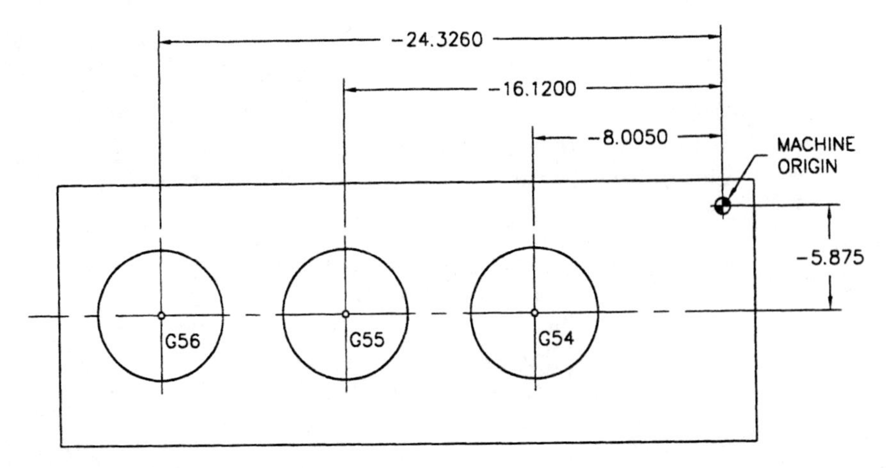

Figure 3–4 Assigning the work coordinates.

According to Figure 3–4, the parts are set in the same line on the machine table; thus all three part origins have the same work coordinates on the Y axis. On the Z axis, the work coordinates have zero value because the home position on the Z axis is kept on the machine origin. For the X axis, each part origin has different coordinate because of different distances from the machine origin.

When using G54 through G59 on most controls, the tool will not collide with the part even if it does not start the cutting cycle from the home position, because these functions position the tool at the home position when they are called up. Then, the tool proceeds toward the part. The G50 and G92 functions are not able to accomplish this. Consequently, the G54 through G59 functions make the machine safer. However, it is best to thoroughly understand the G50 and G92 functions before studying how to preset the coordinate system using the work coordinates. Then it is easier to comprehend how the newer systems work.

Tool Offset Consideration

Tools differ in length as well as diameter, making compensation in slide movements necessary to accommodate the dimensional variations of the tools. This compensation is known as the tool offset. Once the tool offset is established, the slide movement is automatically adjusted according to the value set.

In essence, tool offset is a dimensional value entered to position the tool cutting edge in relation to a programmed point. This point may be the part origin or the part intersection point. According to this, there are two types of offsets:

1. Tool length offsets
2. Geometry offsets

As the name suggests, tool length offsets are used as compensation for tools that differ in length. In the same manner, the geometry offsets are used as compensation for the tools that differ in diameter.

Tool offsets enable using the same part program for any tool length and any tool diameter. This feature is one of the greatest advantages of CNC machines over conventional machines.

Tool offsets have a direct effect on the part size, but they do not affect the part program. This means that the operator may change the offset values a number of times if needed, but different offset values are executed by the same offset number in the program. Thus, the offset may also be described as the operator's way of programming the part.

Tool Length Offsets

When tools are programmed in relation to the part origin, tool length offsets are used to compensate the different tool lengths. Thus, the **tool length offset** may be described as a procedure for correcting the coordinate system preset block in order to achieve the actual tool position in relation to the part origin. This correction is normally accomplished using the offset compensation values rather than changing the values in the program.

On the machining center, tool length offsets are applied in the Z axis only. They are entered by the operator using MDI during the machine setup. The operator must establish the length of each tool according to the part origin. To do so, he or she positions the tool at start point, a position from which the tool starts the cutting cycle. Then the tool is moved to the established zero point level, which may be the part origin in the Z direction or a setting block placed on the part. After touching off the zero position by the tool tip, the operator takes the reading and enters the difference between the programmed and real tool length as the tool offset. He or she repeats the procedure until tool length compensation is established for each tool.

In practice, the tool length compensation is performed by the control in the following simplified manner:

1. The control reads the programmed tool length for a particular tool, say G92 Z10.0.
2. The control checks the buffer for the tool length offset, say 1.0 inch.
3. The control makes a calculation: 10.0 – 1.0 = 9.0 inch.
4. The control assumes that the tool is 9.0 inches long, not 10.0 inches as programmed.
5. In order to reach zero point level, the control lowers the tool for 9 inches since it is 1 inch longer. If the tool is lowered 10.0 inches as programmed, it would collide with the part.

Note that there are different ways of programming the coordinate system preset block. Accordingly, an operator would use the different ways of establishing tool length offsets. For instance, if the tool length is programmed as 15.0, the offset may be –0.5 in order to compensate for the real tool length, which is 15.5 inches. In another case, the tool length may be programmed as 0.0, or not programmed at all. The offset may be –15.5, which is the real tool length. In either case, the tool length offset is used to compensate for the difference between the tool length programmed and the length of the tool used.

On the lathe, tool length offset is applied on both the X and Z axes. For example, the G50 position for one tool is specified as X15.0 Z5.0, meaning the tool tip is 15 inches away (on diameter) from the part centerline and 5.0 inches away from the part face, where the program origin is set. When setting up the machine, the operator sees the following measurement on the screen: X14.9 Z5.1. Thus, there is a difference between the programmed and the real tool length, which is 0.1 inch on the X axis and a 0.1 inch on the Z axis (Figure 3–5).

The G50 position specified in the program will cause the tool to follow the dotted line in Figure 3–6. The resulting part would not be good because it would be 0.1 inch longer (when facing is performed) and 0.1 inch smaller in diameter. To prevent this, the operator must correct the coordinate system. It is not necessary to edit the program and change the G50 values. The flexibility of CNC machines allows compensating for the difference in a faster way, by entering tool offsets values into the tool offset registers. In this case, the particular offset number would be called and a 0.1 is entered to shift the tool upward in a plus-X direction. A –0.1 is entered for the Z axis. Then, the tool shifts toward the part in a minus-Z direction.

When the control starts program execution for the tool shown in Figure 3–5, it reads the programmed coordinate system preset block, G50 X15.0 Z5.0. These

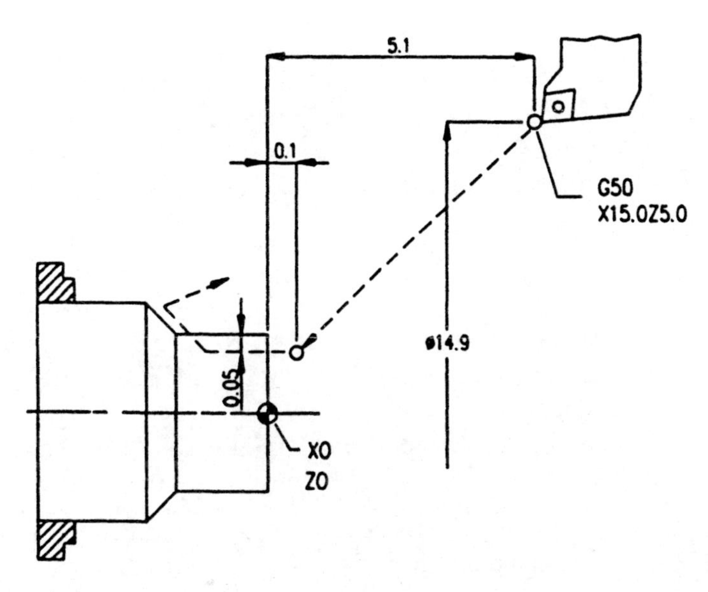

Figure 3–5 Tool offset on the lathe.

values are accepted as the tool distance from the part origin. Then the control reads the block with the tool and tool offset call, T0303. This shifts the tool start position automatically. The shift amount depends on the values stored in tool offset register number 3 for tool number 3. In Figure 3–5, those values would be X0.1 and Z-0.1. The tool length previously accepted through the G50 position has now been corrected. Thus, the part would be machined to proper size, although the tool length as originally programmed was not correct.

When changing the offset values, the operator should be careful about whether he or she is entering an absolute or incremental value. An absolute entry will establish a new offset value, while an incremental value will increase or decrease the offset value previously set. Note that for machines with the left-hand coordinate system, the sign for entering the offset values is reversed.

Geometry Offsets

When milling and turning, although the tool radius center is programmed as the reference point, the point on the tool periphery does the cutting. This point is an r distance away from the tool radius center. Thus, the center of the tool must be shifted away from the part when cutting. This shift amount is known as **geometry offset.** When a geometry offset is called up, the control reads the value entered and shifts the tool away from the part by a specified amount. This feature may be used for the following purposes:

1. To make setup easier when tools differ either from the original tool diameter on the mill or from the tool nose radius on the lathe
2. To affect the part size using the same tool diameter or tool nose radius
3. To achieve the roughing and finishing cuts using the same programmed data

The way of applying the geometry offset is the same either on the lathe or the machining center, as will become clearer later in this book.

Calling and Cancelling Tool Offset

In a lathe program, the T address and a two- or four-digit number are used to call the tool and tool offset. The number of digits may be set by a parameter. The first one or two digits call the tool numbers, which usually range from 1 to 12, while the third and fourth digits call the tool offset numbers, which usually range from 1 to 32 in groups of 8 or 16.

On the machining center, the tool is called by the address T using one- or two-digit numbers, which also may be set by the parameter. It is up to the programmer whether to use a two-digit or four-digit number on the lathe or a one- or two-digit number on the machining center. Using fewer digits saves programming time, but the drawback is that fewer digits allow fewer tools and tool offsets for machining. The offset, usually called by the H address, can be cancelled by the G49 code. The tool numbers usually range from 1 to 32, and the tool offset numbers usually range from 1 to 99.

The T function is used in conjunction with the G00 code in order to generate rapid motion. Following are some examples of calling and cancelling the tool and tool offset:

G00 T0505; (Calling tool number 5 and offset number 5 on the lathe.)
G00 T0500; (Cancelling offset for tool number 5 on the lathe.)
G00 G43 H09; (Calling offset number 9 on the machining center.)
G00 G49; (Cancelling tool offset for the tool presently used on the machining center.)
G00 T03; (Tool selection on some machining centers. Note that tool 3 is now in waiting position, but the change cannot be made by this command.)
G00 T06; (Tool change. The tool in waiting position will be changed for the tool in the spindle.)

When the cutting cycle is finished, the programmer must **cancel the tool offset** in order to return the tool to the start position. For example, T0300; cancels the tool offset for tool number 3. If the offset was not cancelled, the tool would not return to the start position. Then the offset for the next tool would be built up, causing at least a part inaccuracy and possibly a crash.

Some machines allow cancelling the offset by entering T0, and it is wise to use this feature for safety reasons, as illustrated in the following example. Let's say a programmer makes a mistake when cancelling the current tool offset. Instead of T0500, he programs T0100. Consequently, tool T01 rotates into position. If the tool is close to the part or chuck, this may have serious consequences. By entering T0 to cancel the offset, the programmer tells the control to cancel the offset for the tool currently being used. Thus, the program is safer.

On the machining center, the equivalent of T0 is G49. On some machining centers, the tool offset does not even have to be cancelled when the machine origin is used as the tool start position. As a general note, the procedure for calling and cancelling the tool and tool offset must be strictly followed according to the machine's programming manual.

There are two ways of programming the offset call in the program. One way is to program the offset pick-up block in a single command, as illustrated in the following series of instructions:

N10 G50 X10.0 Z3.0 M42; (Coordinate system preset.)
N20 G00 T0101; (Tool and tool offset call on the lathe.)
N30 G97 S550 M03; (Spindle start.)
N40 G00 X3.0 Z0.1; (Approaching the part.)
...; (Machining.)

The code in line 20 calls up the tool and tool offset. After the tool rotates into position, the offset call picks up the values entered into the particular tool offset register. A large value causes the turret to make a noticeable move after the offset reads in. The amount of this move is equal to the offset value. A really large value may cause the tool to interfere with the part or machine when picking up the offset.

The other way of programming the offset pick-up is to combine the offset call and the move command in one block, as illustrated below:

```
N10 G50 X10.0 Z3.0 M42; (Coordinate system preset.)
N20 G00 T0100; (The tool rotates into waiting position.)
N30 G97 S550 M03; (Spindle start.)
N40 G00 X3.0 Z0.1 T0101; (Approaching the part and offset call.)
```

The code in line 20 only calls up the tool to rotate into position; there is no offset call yet. In line 40, the move command is programmed in order to approach the part. The same block calls up the offset. These two commands will produce a smooth motion toward the part. A word of caution: In line 40, the tool in waiting position must be called. In this particular case, it is tool T01. Otherwise, another tool will start rotating into position while the turret is approaching the part.

Tool Offset Adjustment

The correction of the coordinate system may be done any time during machining. It may be achieved by changing the values in the coordinate system preset block, but a faster and safer way is by changing the offset values. For example, take a part the tapping depth of which, on the blueprint, is specified as 1.000 inch. The machine taps the hole to 0.975 inch. The operator should then make the adjustment in size for the next part. The adjustment is to be made in a minus-Z direction. The hole should be cut 0.025 inch deeper; thus the incremental offset adjustment will be –0.025 inch. If the previous offset was 0.020, the control will make a calculation and now show the offset at –0.005.

On the lathe, the part diameter is machined 0.005 inch under size. The offset adjustment in the X axis is needed since the diameter should be fixed. The adjustment is to be made in the plus-X direction to achieve a larger-than-previous diameter. The incremental value is entered as U0.005 using the decimal point format or U50 using the leading zero supression format.

Sometimes it is efficient to perform a facing operation on both sides of the part in one setup. Then the operator can use the same tool with two offsets: one offset to clean up the first face and the other to cut the part to overall length. If doing only a facing operation, the operator simply repeats the program. The block delete code will branch the program when it is needed to distinguish the offsets, as illustrated below:

```
N10 G50 X10.0 Z3.0 S1100 M42; (Coordinate system preset. Speed limit. High
    range.)
N20 G00 T0101; (Tool and tool offset call to cut the first face.)
N30 /G00 T0111; (Tool and tool offset call to cut the other face.)
```

N40 G96 S500 M03; (Spindle start.)
N50 G00 X2.5 Z0; (Rapid to position for the facing cut.)
N60 G01 X-0.064 F0.09 M08; (Facing; the tool is programmed to pass the centerline for a twice tool nose radius value.)
N70 G00 X3.0 Z0.5 M09; (Leaving the part. Coolant stop.)
N80 X10.0 Z3.0 T0100 M05; (Return to home position. Offset cancel.)
N90 M30; (Program stop and rewind.)

Line 20 calls up tool T01 and tool offset 01 to cut the first face. Then, the program execution jumps over line 30, assuming that the block delete switch is on. The execution continues in line 40. When the program is completed, the operator turns the block delete switch off and reruns the program. Then both the N20 and N30 blocks read in. In line 20, tool 01 and offset 01 are picked up. In line 30, the tool does not rotate because the same tool used in line 20 is used again, but offset 01 is changed to offset 11. Line 80 cancels either offset for tool T01.

The control is made in such a way that it executes the last instruction if two pieces of the same kind of information are entered in the program. For example, if the following is programmed,

Y1.0 Y1.5;

the control executes the last value specified as Y1.5. This feature may be used by experienced programmers and operators to solve the problem with the block delete code more effectively by placing it in the middle of the instruction block. Thus, the N20 and N30 blocks in the previous program may be combined in one block, as follows: N20 T0101/T0111;. When the block delete switch is on, the control ignores the information after the slash code and calls up offset 01 to clean the first face. When the block delete switch is off, a complete N20 block reads in, but only the last information is executed, producing offset 11 to cut the other face. The correction of the part size may be accomplished as follows: If the first face is not cleaned up, offset 01 should be adjusted in a minus direction. Then, the tool will cut more. If a change in overall length is needed, offset 11 should be adjusted in a plus or minus direction, depending on whether the part needs to be longer or shorter.

In actual practice, there are many programming tasks that can be made easier using the block delete code, including supplying the coolant, choosing the different speeds and/or feed rates for different materials, calling a subprogram, selecting a particular tool, and repeating the parts. The list is limited only by our imagination.

Methods of Programming the Coordinate System

The tools are set up according to the values programmed in the coordinate system preset block. There are three ways of programming this block:

1. Real values
2. Same imaginary values
3. Approximate real values

It is in the programmer's discretion how to program the coordinate system preset block. Several factors may influence the programming of this block, but one of the most influential is the number of machines per programmer.

Real Values

Programmers using **real values** in the coordinate system preset block are using the tools' real distances from the part origin. Consequently, the tool offsets are zero. When programming, the programmer enters zero values or leaves blanks in the coordinate system preset block. The operator calculates the real distances from the tool tip to the part origin and enters them in the program. This is a very fast and safe way of setting up the tools. Best results are achieved when the same person is both programmer and operator. Following is an example of entering real values in the coordinate system preset block when programming for the lathe:

```
Tool 1
    G50 X0 Z0; (Entering zero values.)
    G00 T0101;
Tool 3
    G50 X_Z_; (Leaving blanks.)
    G00 T0303;
```

The machine indicates an error if the operator does not change the blanks. This is not the case when using zero values, because zero has its own value of zero. Thus, no error is indicated, even if the operator does not change zero values programmed for the real values. This may result in unexpected motions of the tool. Consequently, the method of leaving blanks in the coordinate system preset block is considered safer than programming zero values.

The programming of real values in the coordinate system preset block is widely accepted in the industry. Note that some programmers and the operators on machining centers enter the real tool distance from the part origin as the offset value. Then, in the coordinate system preset block, they enter a zero value or no value at all when using the G54 through G59 functions.

Same Imaginary Values

When using this method, programmers do not use the tools' real distances from the part origin. The values entered in the coordinate system preset block are imaginary numbers suitable for several machines. Tool offsets may be large; it is not unusual if their values are 2, 3, or more inches.

Programming the coordinate system preset block with **same imaginary values** is not considered safe. Even if the offset value is changed by accident, the operator may not notice it because the offsets are normally large. This may have serious consequences. Following is an example of presetting the coordinate system using the same values for all the tools in the program:

Tool 1
```
G92 X20.0 Y12.0 Z10.0;
G00 T01;
```
Tool 5
```
G92 X20.0 Y12.0 Z10.0;
G00 T02;
```

The operator reads the values programmed and calculates the real values for each tool. The difference between the programmed and real values is compensated for through the offsets. It is wise to double-check these calculations because the offsets values may be large. This will ensure that correct offset values are entered. The method of presetting the coordinate system using the same imaginary values for all the tools is mostly used in larger CNC departments where one programmer designs the programs for several machines.

Approximate Real Values

This system involves the programmer entering the approximate tool lengths for all the tools. The programmer enters round numbers that are very close to the real G50 or G92 values. The offsets are not so large, usually up to 0.2 inch. If the operator enters a larger offset value by mistake, it can be noticed easily. When a smaller number that was entered incorrectly goes unnoticed, it should not have serious consequences. Thus, programming the approximate real values is still safer than using the same numbers for all tools in the program. Following is an example of programming the approximate real values in the coordinate system preset block:

Tool 2
```
G50 X15.1 Z4.9; (The real value is X15.127 Z4.880.)
G00 T0202;
```
Tool 3
```
G50 X13.9 Z3.4; (The real value is X13.890 Z3.425.)
G00 T0303;
```

The method of programming the approximate real values is mostly used in smaller shops where the programmer writes the programs for the known tools to be used. This method is very safe because the programmed values in the coordinate system preset block are close to the real values. The difference is usually less than 0.1 inch. If the operator finds that the calculated offset values are larger than usual, he or she can question the calculation.

Setting Up Tools on the Lathe

Setting up the tools is a very important procedure; it affects the sizes of the finished part as wells as the safety of the operator and machine. Tools may be set using a qualified tooling method, a tool preset method, or a touch-off method. For qualified tooling, special tool holders with close tolerances are used. The tool preset method is performed using the tool room apparatus and fixtures in order to duplicate dimensions necessary for setup. Both of these methods save time if properly executed, but experienced operators prefer the touch-off method because of its simplicity and dependability and since the operator controls the procedure. Newer lathes have a measuring arm that swings down. The operator touches the tool to two pads to calculate the offsets.

When setting up tools using the touch-off method, the operator should set the zero reading at the home position. Until the zero reading is changed, any number of tools may be set by repeating the procedure a number of times.

On the lathe, the operator needs to know the screen reading on the Z and X axes as well as the part diameter in order to know where the tool is according to the part origin (Figure 3–6).

In Figure 3–6, the tool is at the programmed G50 position, which is 10.0 inches (on diameter) from the part origin on the X axis. In order to touch off a 3.0 inch part diameter, the tool will travel 7.0 inches, as shown on the screen reading. When adding the screen reading and the part diameter, the result is the value programmed in the coordinate system preset block (7.0 + 3.0 = 10.0). Thus, there is no need for compensation. The offset value is zero.

In reality, a tool of that length might not be available, so another tool must be used. This tool may be 0.5 inch longer. In order to touch off the part, the tool travels only 6.0 inch (on diameter). When adding the screen reading and the part diameter (6 + 3 = 9), the result is the real G50 position for this tool. The difference between the programmed and the real value is 1.0 inch (10 – 9 = 1). The tool must be shifted away from the part in a plus-X direction. Either a change in the coordinate

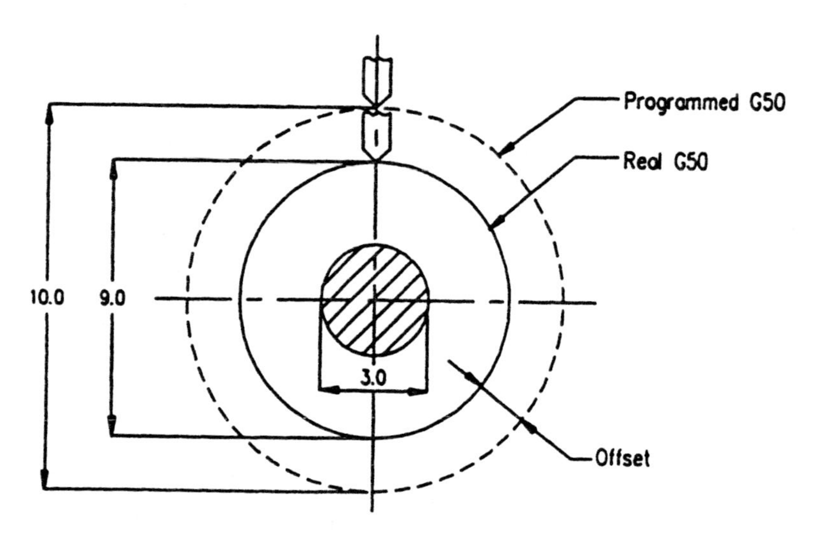

Figure 3–6 Setting up tools on the lathe.

system preset block or compensation through the offset is necessary. If the operator does neither, the part dimensions will be 1.0 inch too small.

The procedure of setting up the tools on the lathe is the same for any of the methods of programming the coordinate system preset block. The only difference is the type of programming used: imaginary tool tip, also known as leading edge programming, or tool nose center. In either case, a widely accepted touch-off method may be used. This is accomplished by manually bringing the particular tool to touch off the part surface. Usually, the finished part face is used to touch off on the Z axis. On the X axis, a light cut is taken on the outside or inside diameter, depending on which tool is to be set.

The Imaginary Tool Tip Method

Whether using imaginary tool tip programming or tool nose center programming, the cutting is performed by the point on the periphery of the tool nose radius. However, these programming methods are distinguished by the reference point used to command the tool along the tool path. In **imaginary tool tip programming,** the **tool reference point** is at the intersection of the lines tangent to the tool nose radius. Although this point actually does not exist on the tool tip, it can be used as the reference point when programming (Figure 3–7).

When setting up tools for imaginary tool tip programming, the reference point and cutting point are on the same line. For example, if the reference point is programmed to a 3.0 diameter, that particular diameter will be machined. Thus, no compensation is needed when cutting along the machine axes (Figure 3–8).

Following is a step-by-step example of setting the tools on the lathe when imaginary tool tip programming is used:

1. Set a zero reading at the home position. Do not change the zero reading while the process of setting up the tools is in progress because the setup of the tools reflects the relation between the part origin and the home position.
2. Face the part to get the part origin on the Z axis. Take a light cut on the part outside and/or inside diameter.

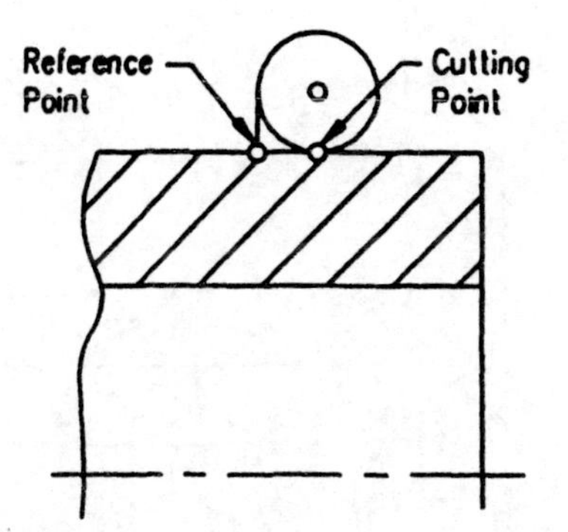

Figure 3–7 Programming by an imaginary tool tip.

Figure 3–8 In imaginary tool tip programming, the reference and cutting point are on the same line, meaning they have the same coordinates.

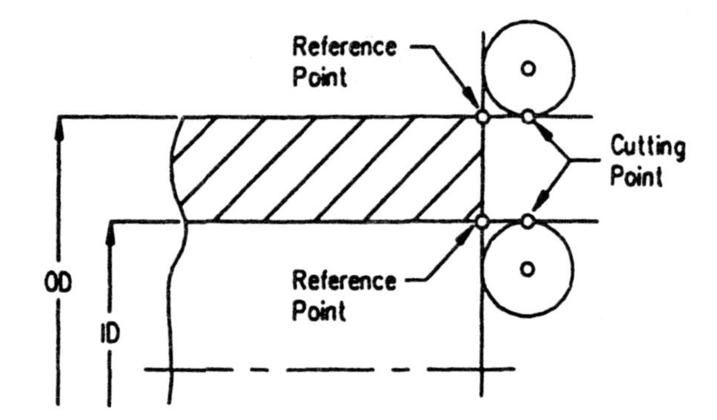

3. Measure the diameters and write them down, such as:

   OD = 6.135
   ID = 4.876

4. Bring the tool and touch off the outside or inside diameter, depending on the purpose of the tool.
5. Take the screen reading on the X axis, which is the U value on the newer machines or the X value on the older machines. Add the reading and the part diameter measured. Thus,

   8.884 (screen reading) + 6.135 (part diameter) = 15.019

   The value 15.019 is the distance from the tool tip to the part origin measured from the home position; consequently, it is the real tool length on the X axis.
6. Check what is programmed in the coordinate system preset block. If using the method of real values, enter the value 15.019 instead of X0 or blank. Then, enter the offset value as zero. If using the same imaginary values method, subtract this number from the value programmed in the coordinate system preset block. Thus,

   15.0 (programmed X value) – 15.019 (real X value) = –0.019 (offset)

7. Enter the difference between the programmed and the real values into the tool offset register as a negative value. This means that the tool has to be moved 0.019 inch in the minus-X direction in order to correct the programmed X value. The tool is shorter than it is supposed to be. The same procedure applies if the method of approximate real values is used in the coordinate system preset block. Thus,

   15.100 (programmed X value) – 15.019 (real X value) = 0.081 (offset)

   Enter the difference into the tool offset register as a positive value. This means that the tool has to be moved 0.081 inch in the plus-X direction in order to correct the programmed X value. This is because the tool is longer than it is supposed to be.

8. For the Z axis, take the reading after touching off the part origin. Then, the reading is the real tool length on the Z axis. This value is entered in the program instead of blanks or zeros if using the real values method. If using one of the other methods, subtract the reading from the programmed Z value. Enter the difference as the offset value, as illustrated in the following examples:

Example 1

5.000 (programmed Z value) – 8.333 (real Z value) = –3.333 (offset)

Enter the offset value into the register as a negative value. It means that the tool has to be moved 3.333 inches in the minus-Z direction in order to correct the programmed Z value. The tool is shorter than it is supposed to be.

Example 2

4.900 (programmed Z value) – 4.895 (real Z value) = 0.005 (offset)

The tool is longer by only 0.005 inch. The reason might be that this tool and the master tool are for the same purpose, say turning or boring. In other words, they are close in length.

On some newer lathes, the process of setting up the tools is performed automatically by the machine **Measure function.** When the part is touched off by a particular tool, the operator selects the offset number on the screen and punches in the part diameter. After pressing the MEASURE button, the machine displays the offset value. This is done quickly and accurately since the control does the screen reading and arithmetic calculations. The operator only enters the size of the part diameter.

The Tool Nose Center Method

Tool nose center programming uses some older programming languages such as APT and Compact II. Newer programming languages also support this method of programming, but programming by the imaginary tool tip is more common, because there is one less calculation, making the imaginary tool tip method a bit more efficient.

Remember that when programming by the imaginary tool tip method, the tool reference point is on the same line as the cutting point when moving the tool parallel to the machine axes. When programming by the tool nose center method, the tool reference point is on the r distance from the part surface, either for diameter or face cutting (Figure 3–9). Thus, it has to be taken into account when setting up the tools.

To calculate the tool length on the X axis, the operator measures the part diameter and adds it to the screen reading in the same way as when setting up the tools for imaginary tool tip programming. However, he or she adds or subtracts a double value of the tool nose radius from the value calculated, as follows:

External tools: D + (2 • TNR)
Internal tools: D – (2 • TNR)

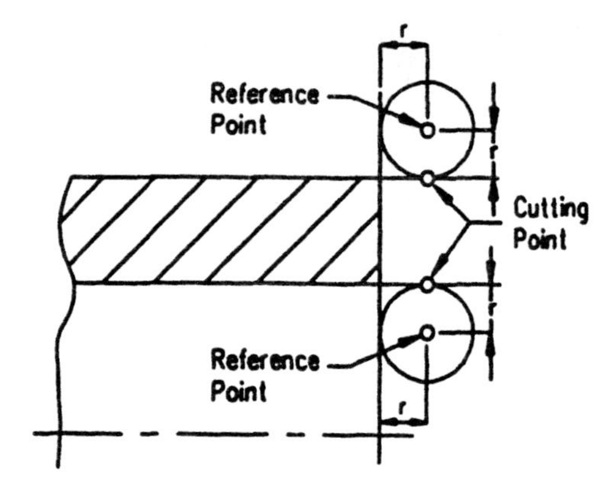

Figure 3–9 When programming by the tool nose center method, the reference point is an r distance from the cutting point.

To calculate the tool length on the Z axis, the operator subtracts value of the tool nose radius from the screen reading, as follows:

For any tool: Reading – TNR

For a better understanding of setting up the tools when programming by the tool nose center method, let's calculate the offsets using the following data:

Outside diameter:

Measure X = 4.135
Reading X = 5.648
Reading Z = 6.145
TNR = 1/32 = 0.03125

Inside diameter:

Measure X = 1.915
Reading X = 7.868
Reading Z = 3.168
TNR = 1/32 = 0.03125

1. The real tool length for the turning tool on the X axis:

   OD measure X + (2 • TNR) + OD reading X = 4.135 + 0.0625 + 5.648 = 9.8455

2. The real tool length for the turning tool on the Z axis:

   Reading Z – TNR = 6.145 – 0.03125 = 6. 1137

3. The real tool length for the boring bar on the X axis:

   ID measure X – (2 • TNR) + ID reading = 1.915 – 0.0625 + 7.868 = 9.7205

4. The real tool length for the boring bar on the Z axis:

   Reading Z – TNR = 3.168 – 0.03125 = 3.1367

The operator should make sure that he or she chooses the insert with the tool nose radius specified in the program. Otherwise, although the calculations are correct, the values entered will not reflect the real tool lengths.

Setting Up Tools on the Machining Center

When programming for a machining center, the programmer usually uses any suitable numbers in the coordinate system preset block. Then, the operator calculates the real tool lengths, which are the distances from the tool tip (when the tool is at the home position) to the part origin. The operator actually corrects the programmed coordinate system by entering the differences between the real and programmed values into the tool offset register. In essence, the programmer enters the imaginary tool lengths into the program and the operator corrects them to the real tool lengths for the particular tools loaded onto the machine.

On machining centers, the tool length may be compensated away from the part using the codes G45 or G43, and toward the part using the codes G46 or G44, depending on a particular control. On the majority of machining center controls the tool offset is cancelled by the G49 code. On some newer controls, the offset may be cancelled by programming T0. Some types of controls allow omitting the tool offset cancel instruction if the tool is programmed to return to the machine origin. When the tool reaches that point, the offset is cancelled automatically. (To be sure if the machine has these capabilities, consult the machine manual.) Following is a description of the procedure of setting up the tools on the machining center (Figure 3–10).

The operator must set up the tools at the part surface chosen by the programmer as the part origin on the Z axis. For oversized parts, rough castings, and forgings, he or she may need to mill off the top of the part manually, taking care of

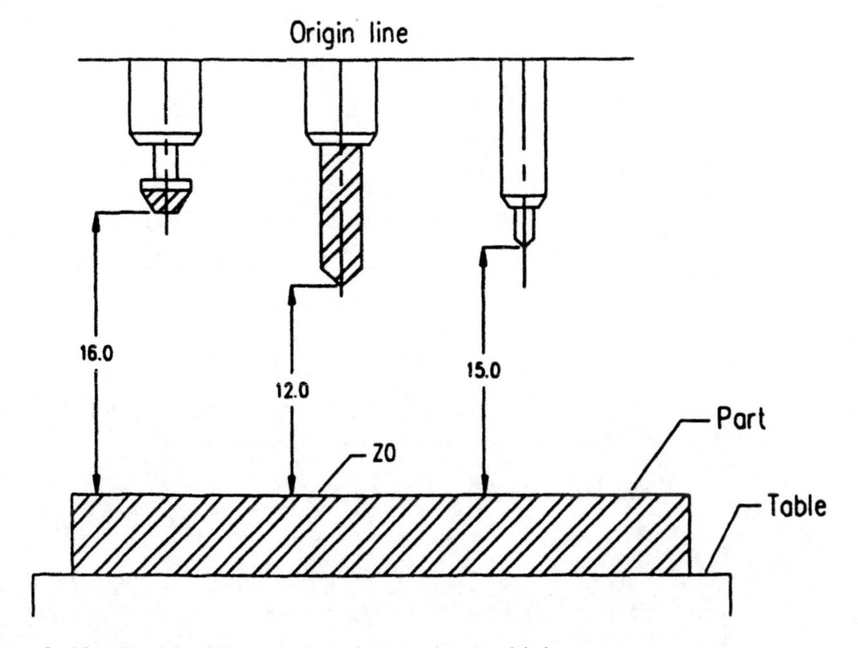

Figure 3–10 Tool length compensation on the machining center.

sizes and allowances for later operations, if any. Then, he or she can proceed in the following way:

1. Position the tool at the home position and set the zero reading on the screen. Do not change the reading until all tools are set up.
2. Lower the tool and touch off the part surface chosen as the part origin on the Z axis.
3. Take the screen reading, which is important for further calculations.

If the Z value in the coordinate system preset block is programmed as 15.0, and if there is a tool of that length, the tool offset will be zero. In Figure 3–10, this tool is a center drill. The programmed first move from the home position to the part origin would be:

In incremental: G91 G43 G00 Z-15.0 H01;
In absolute: G90 G43 G00 Z0 H01;

In both cases, the tool travel is 15 inches because the stored offset value for H01 is zero.

In practice, the tools are longer or shorter than the value programmed in the coordinate system block. When the tool is longer, as is the drill in Figure 3–10, the distance from the tool tip to the part origin is shorter. Then, the difference is calculated as follows:

15.0 (programmed Z value) – 12.0 (reading in Z) = 3.0 (offset H02)

The operator may choose one of the following three ways to correct the coordinate system preset block:

1. Enter the real value instead of blanks or zeroes; i.e., G92 Z12.0;. Then the offset value will be zero.
2. Keep the zero value programmed in the coordinate system preset block. Then the offset value will be –12.0.
3. Keep the value 15.0 programmed in the coordinate system preset block. Then the offset value will be 3.0.

To reach the part origin, the tool will travel 12 inches in either absolute or incremental programming due to the offset value stored into the tool offset register H02. Notice that the offset value is different in each case.

When the tool is shorter than the imaginary tool, as is the countersink in Figure 3–10, the correction of the coordinate system is done in the same manner as before:

15.0 (programmed Z value) – 16.0 (reading in Z) = –1.0 (offset H03)

The same consideration applies to this tool as to the longer tool:

1. Keep the zero value; i.e., G92 Z0;. Then the offset value will be –16.0.
2. Enter the real value instead of blanks or zeroes; i.e., G92 Z16.0;. Then the offset value will be zero.
3. Keep the imaginary value or approximate value of 15.0 programmed in the coordinate system preset block. Then the offset value will be –1.0.

Setup Information

To make machine setup easier and safer, the programmer provides the operator with vital information written on a form known as the **setup sheet.** By reading and understanding this information, the operator gets a clear picture of how the program is designed.

A setup sheet is the link between the programmer and the operator. Any skilled operator should be able to use this sheet to run a program and produce a satisfactory machined part. This is possible even if the program was made years ago or the programmer has left the company, but only if the setup sheet is saved over time.

Any information may be entered in a setup sheet at the programmer's discretion. The sheet is a list of short, concise, and meaningful information, not a short CNC course. The following example shows the optimum amount of information to be written on a setup sheet when programming for the lathe. Notice that some examples are entered.

1. Description
   - Machine number: CINLATHE 2
   - Program number and date
   - Programmer's name (by name or code)
   - Part drawing number: A-00125
   - Operation number: 3
   - Material: type and hardness
   - Stock size
2. Home position on Z: 3.0 inches from Z0 by T05
3. Tooling
   - Face and turn: T01, offset 01, TNR 3/64
   - Drill: T03, offset 03, TNR 0
   - Bore: T05, offset 05, TNR 1/32
   - Internal thread: T07, offset 07, TNR 0
4. Fixtures and gauges
   - Chuck jaws: soft, bored through
   - Special gauges: 1.000–8 UNC male guage

A setup sheet is usually produced as a form. The programmer fills in the blanks when the program is ready. Then, the program, setup sheet, and blueprint are given to the operator, who will set up the machine according to this information. The operator should carefully read the sheet to check if there is enough information to perform the machine setup and proceed with machining. The operator will need to know:

1. Is the program made for one particular operation for the part shown on the blueprint? The program may be for another operation or for a different part.
2. Is the stock size as written on a setup sheet? If not, the program should be adjusted. Stock size changes are a common reason for the program adjustment.

3. Do the tools loaded onto the machine have the same numbers as in the program? If not, a change has to be made in the order of tools already loaded or in the tool numbers programmed. This is a common reason for program changes. Operators prefer to change the tool numbers in the program since it is faster and easier. Then, the operator must not forget to change the tool number at each place where the particular tool appears in the program. The same applies for the tool offset numbers.
4. The operator should check the number of long tools, such as boring bars, drills, and reamers. When more than one long tool is to be used on the lathe, the operator must find a proper way to load all the tools. One popular way is to load the tools using even numbers for the long tools and odd numbers for the short tools such as, T01-turning tool, T02-drill, T03-external threading tool, T04-boring bar, and T06-internal grooving tool. This configuration of the tools will make sure that no tool interferes with the chuck while another is cutting.
5. The operator should pay attention to the tool radius to be used. If a wrong value is entered, the geometry offset will not be correct. This causes part inaccuracy.
6. Some part sizes are dependent on special fixtures and gauges. In order to achieve the expected result, the operator should use those specified by the programmer on the setup sheet.

In general, the operator should carefully observe any information written on a setup sheet. He or she should also look for any improvement that can be made either in the program or setup procedure and pass it on to the programmer. When the same job is run again, the operator will achieve better results.

Screen Reading

Screen reading is an important skill that can be developed only through experience. This skill is especially needed when proving the program machining the first part. Many serious problems can be avoided through screen reading.

There are different screens on the monitor of a CNC machine. Some of them are presented below:

- Program screen
- Position screen
- Offset screen
- Work zero screen
- Parameter screen

Each of these screens provides valuable information about the status of the machine tool, but the most important screens for the operator are the Offset and Position screens. If the machine has work shift capabilities, the Work Shift screen is also important to performing machine setup.

Offset Screen

The lathe offset screen is similar on all lathes. Also, the machining center offset screen is similar on all machining centers. Following is the first page of the offset screen taken from one popular machining center:

TOOL OFFSET

	LENGTH		RADIUS	
NO.	**GEOMETRY**	**WEAR**	**GEOMETRY**	**WEAR**
001	–19.1900	0.0000	0.0000	0.0000
002	–20.1300	–0.0100	0.0000	0.0000
003	–17.9000	0.0000	0.0000	0.0000
004	–20.7300	0.0000	0.0000	0.0000
005	–19.8820	0.0000	0.0000	0.0000
...	...	...	...	...
015	–12.5770	0.0000	0.5000	0.0050
016	–10.2250	0.0000	0.3750	0.0000

At the top is the name of the screen. The screen is divided into two parts: Length and Radius. These are short for tool length offset and tool geometry offset. The tool offset numbers are listed from 1 to 16. GEOMETRY shows the offset entered when the setup was made, and WEAR shows the tool offset adjustment during machining.

The geometry offset number 002 has the value of –20.1300, and an incremental offset adjustment of –0.0100. Thus, the total value of this offset is –20.1310. Note that the geometry offset may also be adjusted in incremental mode. Then the wear offset would be zero. This means that if an offset adjustment is needed, it should be made in geometry or wear, but not in both. (It is better to use WEAR.)

Offset number 015 has a value of 0.5000 in RADIUS offset and 0.0050 in WEAR. This means that this program uses a 0.5 inch tool radius for geometry offset. The offset was adjusted for 0.005 inch during machining. Also, the adjustment can be made in GEOMETRY by entering 0.5050. Then the value in WEAR should be changed to zero. Again, better to use WEAR.

Work Zero Offset Screen

All newer machines have the **Work Zero Offset function,** which speeds and simplifies setup. Following is a page from the Work Zero Offset screen of a popular machining center:

WORK ZERO OFFSET

NO.	**00 (COMMON)**	**NO. 02 (G55)**	**NO. 04 (G57)**	**NO. 06 (G59)**
X	0.0000	X –27.0850	X 0.0000	X 0.0000
Y	0.0000	Y –9.1635	Y 0.0000	Y 0.0000
Z	–0.0100	Z 0.0000	Z 0.0000	Z 0.0000
A	0.0000	A 0.0000	A 0.0000	A 0.0000

	NO. 01 (G54)	NO. 03 (G56)	NO. 05 (G58)
X	–40.4850	X 0.0000	X 0.0000
Y	–9.1635	Y 0.0000	Y 0.0000
Z	0.0000	Z 0.0000	Z 0.0000
A	0.0000	A 0.0000	A 0.0000

INCH

When reading data on this screen, the operator can conclude how the machine is set up. G54 and G55 are used as work zero offsets. This may mean that two parts can be machined simultaneously. The X and Y values tell the distance from the part origin to the machine origin. Notice that the Y value is the same in both the G54 and G55 work zero offsets, while the X value is different. This may mean that two parts are set beside each other when machining.

The values in the COMMON offset affect all work offsets. This means that the value of –0.0100 inch entered for the Z axis in the COMMON offset will shift the tool start point in all work offsets. This is useful when several work offsets are used, because it is faster and safer to make one change in the COMMON offset than to make the change in each work offset.

Position Screen

The position screens are very similar on all of the controls. Following is the position screen taken from one popular type of lathe:

POSITION

RELATIVE	**ABSOLUTE**
U 0.0000	X 15.5720
W 0.0000	Z 5.0210
MACHINE	**DISTANCE TO GO**
X 0.000	X 0.0000
Z –749.300	Z 0.0000

The values in the **Relative position** express the relative tool distance from the home position. This distance is shown in the relative or incremental coordinates, U and W. When reading the values in the Relative position, it is not possible to know directly how far the tool is from the part origin, just how far it is from the home position. The sign of the coordinates is zero or negative because the tool cannot move farther than the machine origin. Thus, at present, the values are zero, so the tool is at the home position. The values in the Relative position are normally used when setting up the tools in order to find the real tool distances from the part origin.

The values in the **Absolute** position express the absolute tool distance from the part origin. This distance is shown in absolute coordinates, X and Z. The sign may be positive or negative, depending on the quadrants in which the tool is moving. This is an important piece of information for the operator when machining, since any value in the Absolute position is directly related to the part.

In the screen reading shown, the values in the Absolute position are the values programmed in the coordinate system preset block for the last tool. Thus, when this particular tool is at the home position, its distance from the part origin is 15.5720 inches on the X axis and 5.0210 on the Z axis. Following are some examples of reading the values in the Absolute position when the tool is cutting the part:

Example 1

X 3.0000 (The X value is decreasing.)
Z 0.0000 (No change.)

When the X value decreases and the Z value does not change, the machine may be face or shoulder cutting.

Example 2

X 0.0000 (No change.)
Z 0.1229 (The Z value is increasing in the plus direction.)

When the X value does not change from zero, and the Z value increases in the plus direction, it may be that the drill is being moved out from hole.

Example 3

X 2.5000 (No change.)
Z –1.8750 (The Z value is increasing in a minus direction.)

When the X value does not change from a particular diameter, and the Z value increases in the minus direction, it may represent any cutting operation along the Z axis, such as turning, boring, or threading.

Example 4

X 2.9999 (The X value is increasing.)
Z –0.5000 (The Z value is increasing in the minus direction.)

When both values increase, it may be taper, chamfer, or taper thread cutting. It may also be arc cutting.

The values in the **Machine position** express the absolute tool distance from the machine origin. As seen from the screen, the tool is at the machine origin in the X direction, and –749.300 mm from the machine origin in the Z direction. This distance can not be expressed with a plus sign. When the tool is at the home position, the values in the Machine position show the distance from the home position to the machine origin. This is not possible when reading the values in the Relative position, since its value is always zero when the tool is at the home position. There are three decimal places in the Machine position; thus this distance is expressed in metric units.

The **Distance to Go** shows the amount of the tool travel needed to reach the programmed destination point. For example, the instruction is programmed as: Z-3.0. When machining, the DISTANCE TO GO value on the screen may be:

X 0.0000
Z –0.5000

This means that the tool will travel another 0.5 inch in the minus-Z direction in order to finish the cutting as per programmed instruction. If the Distance to Go value is greater than the real tool distance from the chuck, the operator should stop program execution. Otherwise, the tool will cut into the chuck. The problem may be that the part is not sticking out far enough or that the tool is not set up properly.

Any information in the Position screen is important when setting up and machining, but the most important are the Absolute position and Distance to Go. They are of great help to the operator. When the screen information is understood properly, the operator always knows where the tool is according to the part origin. He or she also knows how far it has to travel to reach the programmed destination.

Summary

The home position may be set at any convenient point inside the machine electronic and mechanical limits. All the tools have to be clear from the part when the tool changes occur. On some machines the tool change position is dictated by the machine design, especially for mills.

The home position is established according to the machine origin, which is set by the machine tool builder and normally cannot be changed. However, it is also related to the part origin.

The purpose of the coordinate system preset is to tell the control how far the tool is from the part origin according to each machine axis. Some newer machines use the work coordinate to establish the coordinate system. These are programmed using the G54 through G59 functions. These functions are very effective when several part origin points need to be established in one program.

Tool offset is a dimensional value entered to position the tool cutting edge in relation to the programmed point. This point may be the part origin or the part intersection point. Accordingly, there are two types of offsets, tool length offsets and geometry offsets. Tool length offsets are used as compensation for tools that differ in length. Geometry offsets are used as compensation for the tools that differ in diameter. Tool offsets enable using the same part program for any tool length and any tool diameter. This feature is one of the greatest advantages of CNC machines over conventional machines.

Tools may be set using a qualified tooling method, a tool preset method, or a touch-off method. For qualified tooling, special tool holders with close tolerances are used. The tool preset method is performed using the tool room apparatus and fixtures in order to duplicate dimensions necessary for setup. Both of these methods save time if properly executed, but experienced operators prefer the touch-off method because of its simplicity and dependability and because the operator controls the procedure. Newer lathes have a measuring arm that swings down; the operator touches the tool to two pads and calculates the offsets.

To make machine setup easier and safer, the programmer provides the operator with vital information written on a form known as the setup sheet. A setup sheet is a list of short, concise, and meaningful information. By reading and understanding this information, the operator gets a clear picture of how the program is designed. Therefore, a setup sheet is the link between the programmer and the operator.

Screen reading is an important skill that can be developed only through experience. This skill is especially needed when proving the program while machining the first part. Many serious problems can be avoided through screen reading.

Key Terms

Absolute position
Distance to Go
fixture offsets
geometry offsets
home position
imaginary tool tip programming
Machine position
manual data input (MDI)
master tool
Measure function
presetting the register
real values
Relative position
screen reading
setup sheet
tool length offset
tool nose center programming
tool reference point
work coordinates
Work Zero Offset function
Zero Return function

Self-Test

Answers are in Appendix E.

1. The ______________ is the point from which the tools start program execution and to which they return.
2. ______________ is used to position the tool to machine zero and afterwards to home position.
3. ______________ is the tool used to set the home position.
4. The G54, G55, G56, G57, G58, and G59 are the ______________ that are used to establish several part origins in one program.
5. Programming the coordinate system preset block is known as ______________.
6. The operator enters work coordinates using ______________.
7. The ______________ are used to compensate for the tools that differ in length.
8. The ______________ are used to compensate for the tools that differ in diameter.

9. A ______________ is used to command the tool along the tool path.
10. In ______________, the tool reference point is at the intersection of the lines tangent to the tool nose radius.
11. In ______________, the tool reference point is on the r distance from the part surface when moving the tool parallel to the machine axes.
12. On some newer machines, ______________ is used for automatic setup of tools.
13. ______________ is a list of short, concise, and meaningful information written from the programmer to the operator.
14. On some newer machines, ______________ is used in conjunction with work coordinates G54 through G59 to simplify setup.
15. The values in the ______________ express the relative tool distance from the home position.
16. The values in the ______________ express the absolute tool distance from the part origin.
17. The values in the ______________ express the absolute tool distance from the machine origin.
18. The ______________ shows the amount of the tool travel needed to reach the programmed destination point.

Relating the Concepts

No answers are suggested.

1. Prior to setting up the tools, at which place should you set a zero reading? Why it is important not to change this reading until the setup is complete?
2. The screen reading is 9.100
   The part OD is 3.350 inches
   TNR is 1/32 inch
   Given this information, what is the real tool length
   (a) When using the imaginary tool tip programming?
   (b) When using tool nose center programming?
3. The screen reading is 8.850
   The part ID is 4.75 inches
   TNR is 3/64
   Given this information, what is the real tool length
   (1) When using the imaginary tool tip programming?
   (2) When using tool nose center programming?
4. The screen reading is X10.550 Z3.150
   The part OD is 5.140 inches
   The programmed G50 is: X15.0 Z3.0
   TNR is 3/64

Given this information, calculate the offset values. Is the tool shorter or longer than it is supposed to be? Explain.

5. The screen reading is 11.500
   The programmed G92 is: Z14.0
   Given this information, calculate the offset values. Is the tool longer or shorter than it is supposed to be? Explain.
6. The blueprint size of the hole depth is 1.50 inch
   The depth of the drilled hole is 1.60 inch
   Given this information, which axis should be offset? In which direction should the offset be applied? How much should the offset be?
7. The blueprint size of the shoulder diameter is 3.750 inches
   The shoulder is machined to a 3.755 inches
   Given this information, which axis should be offset? In which direction should the offset be applied? How much should the offset be?

4 Manual Tool Radius Compensation

Key Concepts

Tool Radius Compensation Consideration

Calculating the Intersection Point
- Using the Pythagorean Theorem
- Using Trigonometric Functions
- Relationships between Angles

Compensation by Imaginary Tool Tip

Compensation by Tool Radius Center

Tool Radius Compensation Consideration

The programmed point on the part is the *command point.* It is the destination point of the tool. The point on the tool that is used for programming is the *tool reference point.* These two points may or may not coincide, depending on the type of tool used and the machining operation performed.

When drilling, tapping, reaming, countersinking, or boring on the machining center, the tool is programmed to the position of the hole or bore center—this is the command point. The tool radius center is used to position the tool at the tool reference point. Thus, for hole or boring operations on the machining center, the command and reference points coincide [position (a) on Figure 4–1].

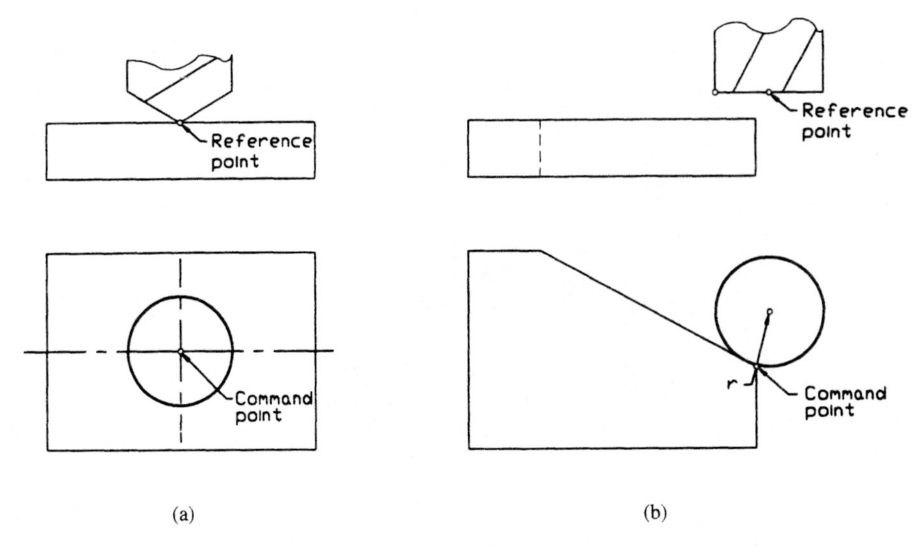

Figure 4–1 The command and reference points for machining center applications.

When milling a contour, the tool radius center is also used as the reference point, but the part is actually cut by the point on the cutter periphery [position (b) on Figure 4–1]. This point is an r distance from the tool center. This means that the programmer should shift the tool center away from the part in order to perform the cutting by the tool cutting edge. The shift amount depends on the part geometry and tool radius. This technique is known as **tool radius compensation** or **cutter radius compensation.** If the programmer does not enter this information in the program, the result is an improperly machined part.

When considering a lathe tool, there is also a smaller or bigger radius on its tip. It is made that way to obtain a stronger and more durable tool tip (Figure 4–2). Consequently, tool radius compensation should also be taken into account when programming the contour on the lathe.

The tool nose center is not the only reference point that can be used for programming contours. On the lathe tool there is a point known as the *imaginary tool tip,* which is at the intersection of the lines tangent to the tool nose radius (Figure 4–3). In reality, this point does not exist on the lathe tool. It is only used as the reference point when programming and setting up the tool.

Programmers must apply tool radius compensation when programming a contour. This can be achieved through the use of the built-in machine software known as *automatic tool radius compensation.* If available, the programmer can use part programming software. When there is no software, he or she must calculate the tool radius compensation in an operation known as *manual tool radius compensation.*

Expertise in manual tool radius compensation is necessary even when using part programming software. This is especially true when adjusting and editing the program on the machine.

Figure 4–2 The lathe tool tip has a radius that makes it more durable. (Courtesy Valenite, Inc.)

Before presenting examples of manual part programming, we will prepare the formulas for some characteristic intersections that are often seen on parts being machined. The principles of plane geometry and trigonometry will be used to create these formulas.

Calculating the Intersection Point

The **intersection point** is the point where two lines or two part surfaces meet. Most of the time, the blueprint does not provide enough details to allow the programmer to program the intersection point directly. Then, the programmer must calculate the coordinates for the intersection point. Usually, he or she starts by drawing a sketch

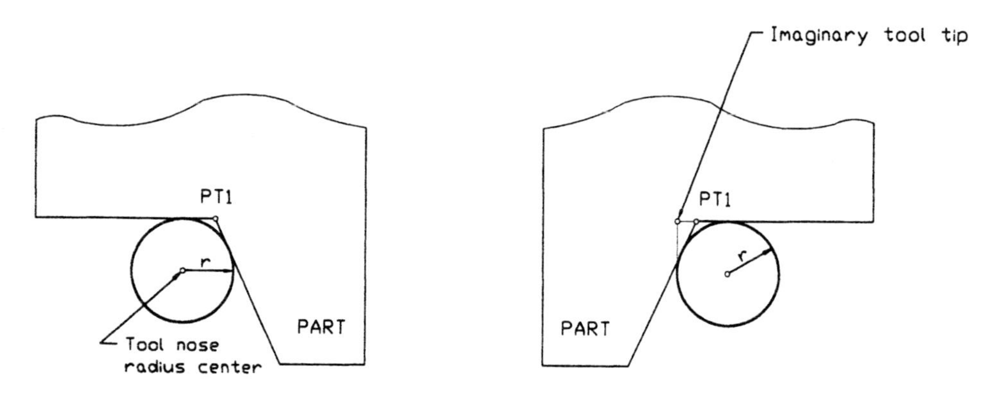

Figure 4–3 On the lathe tool, either the tool nose center or imaginary tool tip can be used as the reference point.

of the intersection point using an enlarged view and right triangles. As needed, the programmer draws the triangles in several positions and then uses them to make multiple calculations until he or she obtains the necessary information. Depending on the specific information sought, the programmer may apply the Pythagorean theorem or trigonometric functions to a particular right triangle.

Using the Pythagorean Theorem

Right triangles contain one angle of 90 degrees (a right angle). In a right triangle there are two sides and a hypotenuse, which is the longest side. The Pythagorean theorem is used when the length of two sides of a triangle are known and the length of a third one is needed (Figure 4–4).

The triangle may be drawn in any convenient way, but it must be a right triangle. Any character may be assigned to the sides or the hypotenuse. We are not limited to *A, B,* and *C* or *a, b,* and *c* (shown in Figure 4–4), even though these are used most often. Following are some examples of applying the Pythagorean theorem when calculating intersection points.

The intersection shown in Figure 4–5 is frequently found on parts machined on both the lathe and the machining center. Using R = 1 and b = 0.6, side *a* is calculated as:

$$a = \sqrt{R^2 - b^2} = \sqrt{1.0^2 - 0.6^2} = 0.8$$

The intersection shown in Figure 4–6, which is known as a *step,* is also characteristic of many parts being machined. The Pythagorean theorem is used to find the value of a after substituting R = 1 and h = 0.5:

$$a = \sqrt{R^2 - (R - h)^2} = \sqrt{1.0^2 - (1.0 - 0.5)^2} = 0.866$$

When there are multiple operations in one formula, the operations inside the brackets must be performed first. In this particular case, the subtraction is performed before the square operation. This is known as the *order of priority of arithmetic operations.*

Using Trigonometric Functions

Sometimes the Pythagorean theorem is not the shortest way to calculate the point because one of the facts given is given as an angle. Then the programmer can use trigonometric functions to calculate the unknown values by knowing two pieces of

Figure 4–4 Right triangle.

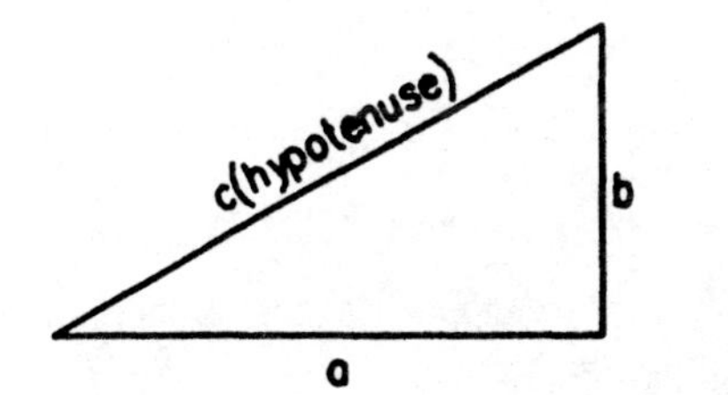

$$c = \sqrt{a^2 + b^2}$$

$$a = \sqrt{c^2 - b^2}$$

$$b = \sqrt{c^2 - a^2}$$

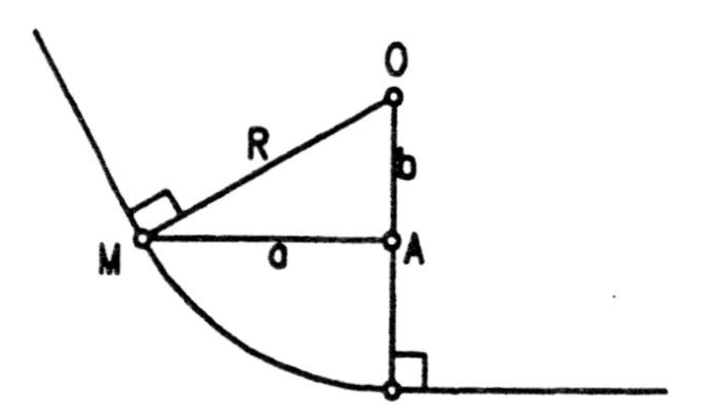

Figure 4–5 Calculating data for the taper and arc intersection.

information about the triangle: either an angle and the length of one side, or the lengths of two sides.

There are three sides in a right triangle that are described with respect to a given angle: the opposite, the adjacent, and the hypotenuse (Figure 4–7). The **hypotenuse** is always the longest side of the triangle. It lies directly across from the right angle. The **opposite side** lies directly across from the known angle. The **adjacent side** lies next to the known angle.

The most common trigonometric functions are sine (sin), cosine (cos), tangent (tan), and cotangent (ctg). The sine and cosine functions need information about one side and the hypotenuse, while the tangent and cotangent functions need information about two sides. The sine and tangent functions deal with the opposite side, while cosine and cotangent deal with the adjacent side (Figure 4–7).

Following are some typical examples of using trigonometric functions when calculating the point of intersection. This calculation is routine for people working on machining centers because there are so many jobs requiring various dimensions for bolt circle hole patterns. These dimensions are not usually given on the blueprint, and often have to be calculated when programming.

Calculating the chord involves several steps:

Step 1. Find the angle between two holes [position (a)] in Figure 4–8. To find an angle, divide a full circle by the number of holes:

$$360 / 8 = 45$$

Step 2. Find the angle between the line perpendicular to the chord and the radius [position (b)]. The line perpendicular to the chord and radius actually splits a 45-degree angle. Thus,

$$45 / 2 = 22.5$$

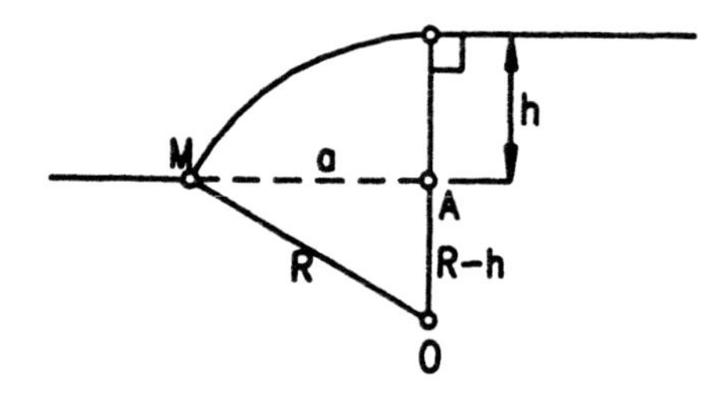

Figure 4–6 Calculating data for an arc and line intersection.

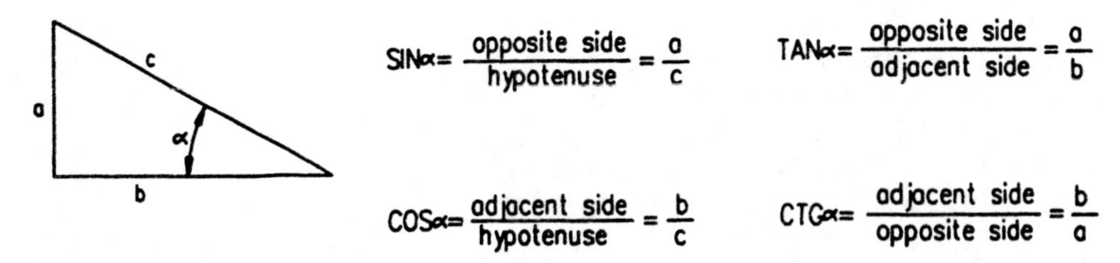

Figure 4–7 Trigonometric functions.

Step 3. Calculate the length of half of a chord from the right triangle [position (c)]. Use the sine function since the chord is a side opposite to an angle; the hypotenuse is known as the radius value. Thus,

$$\sin 22.5 = a / 5.0$$
$$a = 5.0 \bullet \sin 22.5 = 1.913$$

Step 4. Multiply this amount by 2. The length of the chord is 3.826 inches.

Notice the importance of drawing a sketch on an enlarged view in order to calculate the necessary information. Often the blueprint does not provide all of the information needed for programming.

Step 1. The following equations may be derived from the triangle 0-M-1 shown in Figure 4–9:

$$\cos 45 = X / R$$
$$\sin 45 = Y / R$$

Step 2. Solve the equations. The results for X and Y should be:

$$X = R \bullet \cos 45 = 1.0 \bullet 0.707 = 0.707$$
$$Y = R \bullet \sin 45 = 1.0 \bullet 0.707 = 0.707$$

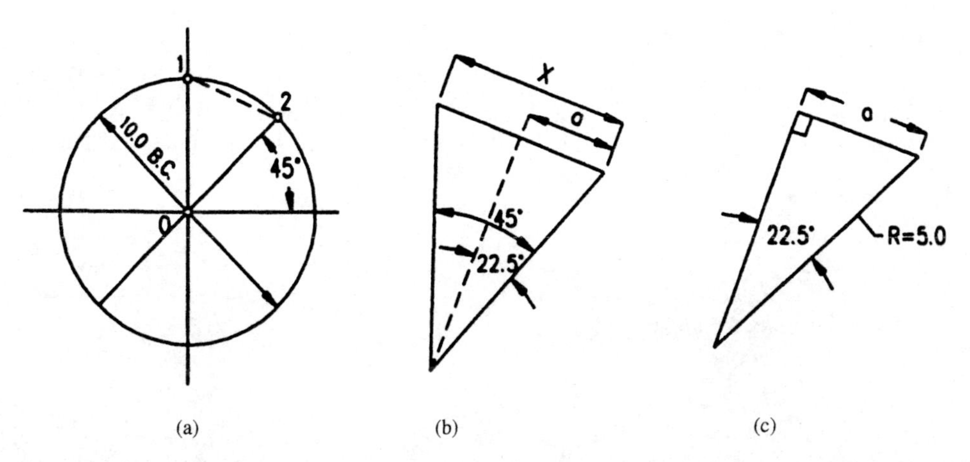

Figure 4–8 Calculating the chord between two holes for an eight-hole pattern on a 10 inch bolt circle.

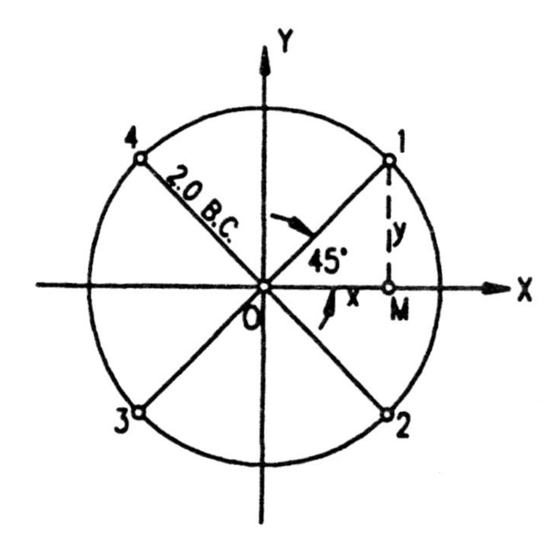

Figure 4–9 Calculating hole center coordinates for a four-hole pattern on a 2 inch bolt circle.

Step 3. Now calculate the hole coordinates. Actually, they are all the same; only the sign is different, depending on the quadrants.

Hole 1	X1 = 0.707
	Y1 = 0.707
Hole 2	X2 = 0.707
	Y2 = –0.707
Hole 3	X3 = –0.707
	Y3 = –0.707
Hole 4	X4 = –0.707
	Y4 = 0.707

Step 1. Obtain the values for a and b in Figure 4–10 using the following equations:

$$\sin 30 = a / R$$
$$\cos 30 = b / R$$

Step 2. After solving the equations using a 0.5 inch radius, you should have the following values for a and b:

$$a = R \cdot \sin 30 = 0.5 \cdot \sin 30 = 0.5 \cdot 0.5 = 0.25$$
$$b = R \cdot \cos 30 = 0.5 \cdot \cos 30 = 0.5 \cdot 0.866 = 0.433$$

Figure 4–10 Calculating data for the arc and taper intersection.

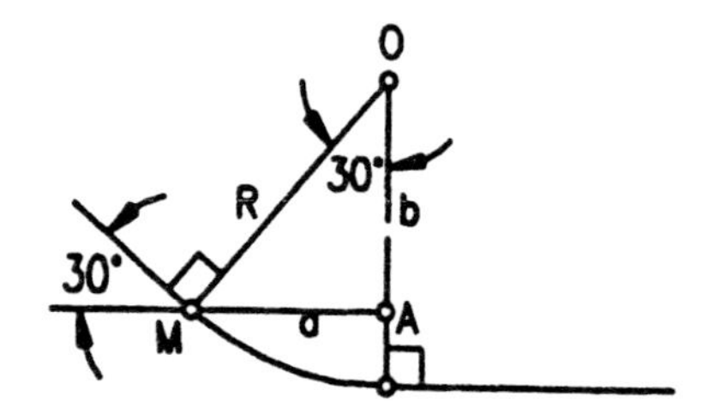

Relationships between Angles

The relationships between the angles enable the programmer and the operator to convert data from a blueprint into the information needed for programming. It helps to figure out which laws of mathematics should be used to obtain the programmable part dimensions.

There are several relationships between angles. These relationships are used when calculating data for the points of intersection (Figure 4–11).

If two lines that make an angle are extended, an equal angle is created [position (a) in Figure 4–11].

If the line is drawn parallel to either line of the angle, an equal angle is created at point A or point 0, as shown at position (b). Also, by knowing an angle α, a programmer can calculate its complementary angle, which is $90 - \alpha$.

At position (c) there is an angle made by the lines m and p. If perpendicular lines are drawn from point 0 to the lines of this angle, a new angle is created. The line that connects one angle origin point with another splits both angles. This relation is especially useful when calculating data for an arc segment.

As per position (d), if one angle is known, an equal angle may be created by drawing the lines perpendicular to the lines which make the first angle. For example, if the angle at point A is known, an equal angle may be created at point B by drawing perpendicular lines from point B to the lines of the angle at point A. The same rule applies if the angle at point B is known and the angle at point A is to be created.

In the following example, the relations between angles are used to calculate the tool radius compensation for a 1-inch-diameter end mill.

As in Figure 4–12, position (a), the angle of the slope is 30 degrees. Using the relations between the angles, the same angle is created at the tool radius center. This is illustrated at position (b). When this angle is split, a 15-degree angle is created, as well as right triangle M-0-A, position (c). From this triangle, the compensation value is calculated:

$$\tan 15 = Xc / r$$
$$Xc = r \cdot \tan 15 = 0.5 \cdot 0.268 = 0.134$$

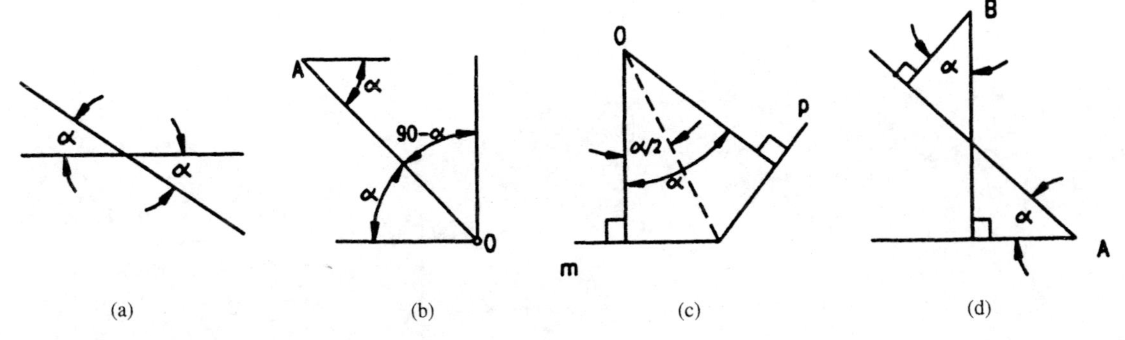

Figure 4–11 Relationships between angles.

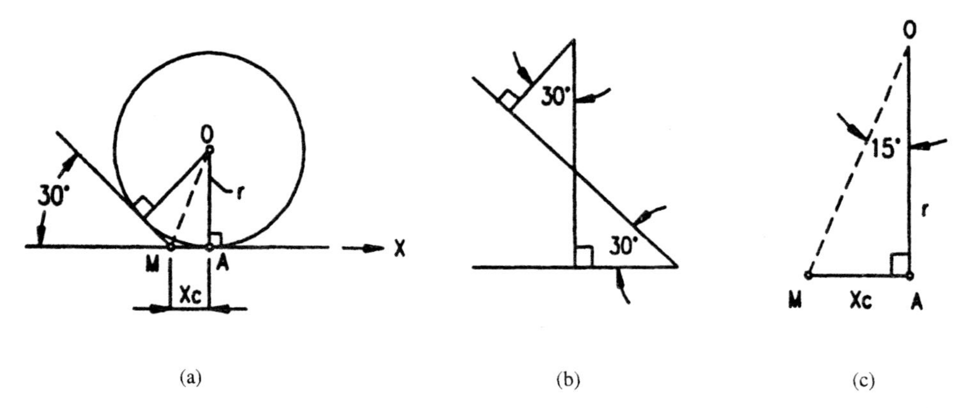

Figure 4–12 Calculating the tool radius compensation using the relationships between the angles.

The value of 0.134 is the tool radius compensation for a 1.0 inch end mill (or 0.5 inch tool radius) when cutting a 30-degree slope. When programming the coordinates of point M, this value is subtracted from the point's X coordinate. For instance, if the X coordinate of point A is 1.0, the tool center, which is used as the reference point, would be programmed to cut up to a 0.866 length on the X axis. Thus, the tool is held for the amount of tool radius compensation. This would ensure that overcutting of the slope does not occur.

Compensation by Imaginary Tool Tip

The imaginary tool tip may be used as the reference point in programming: In such cases, when the tool is cutting parallel to the machine axes, there is no need for calculating a compensation amount. However, when a taper, arc, or chamfer is cut, the tool nose radius compensation has to be taken into account (Figure 4–13).

As in Figure 4–13, the tool nose is held or pushed when chamfering. This enables placing the tool nose perpendicular to both surfaces. At point A, the imaginary tool tip is "held" (programmed to stop before a point) on the X axis. This is because of the difference in the X coordinates between the reference point and the part intersection point. At point B, the imaginary tool tip is "pushed" (programmed to pass) the intersection point. The push is made on the Z axis. If the tool is not held on the X axis at point A, the part will be cut on a larger diameter. If the tool is not pushed on the Z axis at point B, the chamfer length will be too small. Therefore, the compensation amount has to be taken into account when programming these points.

At points C and D, the tool moves parallel to the machine axes. There is no need for compensation, because the reference point and the part intersection point are the same.

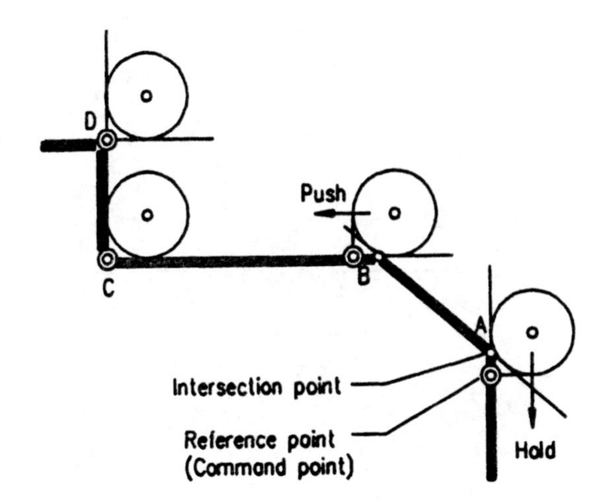

Figure 4–13 When programming by the imaginary tool tip, the compensation is not needed when cutting parallel to the machine axes.

Three cases of a two-line intersection are illustrated in Figures 4–14, 4–15, and 4–16.

As seen in Figure 4–14, the compensation on the lathe Z axis is equal to zero. This is because the intersection point and the tool reference point are on the same line on the Z axis, meaning that they have the same coordinates. In general, if a programmer is using imaginary tool tip programming and the tool is cutting parallel to one axis, the compensation for the other axis is equal to zero.

The compensation for the lathe X axis is calculated according to an angle and the size of the tool nose radius, using trigonometric functions. The result is added or subtracted from the point's X coordinate, depending on the side from which the tool is coming to the intersection and whether an outside or inside diameter is being cut. When cutting a chamfer on an outside diameter, the programmer holds down the imaginary tool tip for the amount of compensation. Thus, the compensation value is subtracted from the point's X coordinate. The opposite happens when the programmer intends to cut a chamfer on an inside diameter. Then, he or she pushes the tool start point upwards on the X axis by adding the amount of compensation to the point's X coordinate.

In Figure 4–15, the compensation for the lathe X axis is equal to zero. This is because the intersection point and the tool reference point are on the same line on the Z axis, meaning that they have the same coordinates. Thus, no compensation is needed.

The compensation for the lathe Z axis has to be calculated. The result is added or subtracted from the point's Z coordinate, depending on the side from which the tool is coming to the intersection. For normal turning or boring operations, the tool is pushed on the Z axis, meaning that the compensation is added to the Z coordinate.

When neither line is parallel to the machine axis, as in Figure 4–16, the intersection is made from two angled surfaces, and the compensation for both axes has to be calculated. When using the presented formulas, the programmer should make the calculations according to the order of priority of arithmetic operations. In this particular case, the following order of priority should be used:

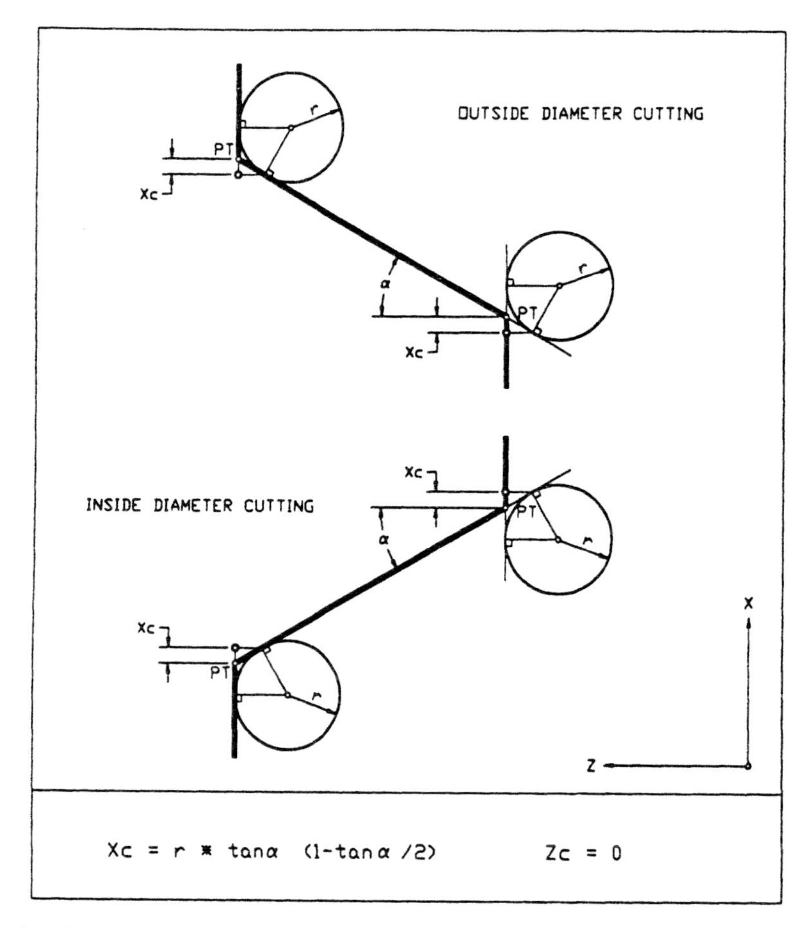

Figure 4–14 Line parallel to the lathe X axis.

- The angles are added or subtracted.
- The result is divided by 2 to get a half-angle.
- The value for sine or cosine is multiplied by the tool nose radius (r).

The following is an example for calculating data for a two-line intersection. For the part shown in Figure 4–17, position (a), calculate the coordinates of points A and B with tool radius compensation when using 1/32 tool nose radius.

Solution: At points A and B, the cutter must be placed perpendicularly to both surfaces [position (b)]. Using the formulas presented earlier, the following calculations can be made:

Point A:

Tool radius compensation on the X axis:

$$Xc = r \cdot \tan \alpha(1 - \tan \alpha / 2) = 0.032 \cdot \tan 30 \cdot (1 - \tan 15) = 0.0132$$

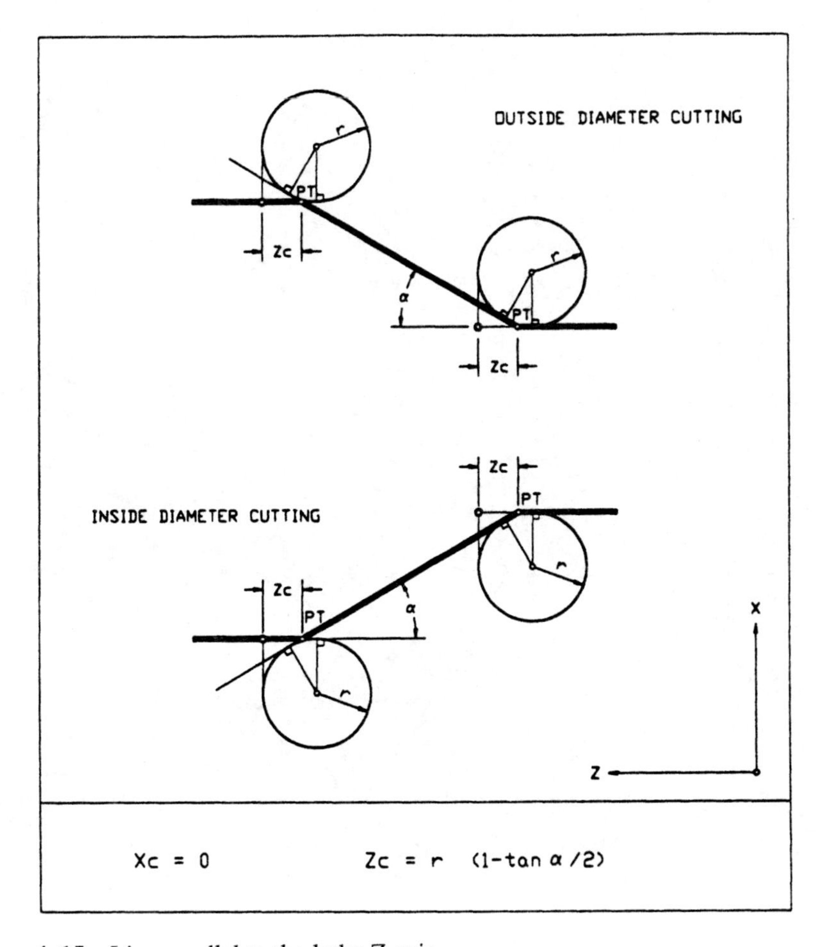

Figure 4–15 Line parallel to the lathe Z axis.

Tool radius compensation on the Z axis:

$$Zc = 0$$

Coordinates of point A without compensation:

$$XA = 3.00 - 2 \cdot (0.5 \cdot \tan 30) = 2.423$$
$$ZA = 0$$

Coordinates of point A with compensation:

$$XA = 2.423 - (2 \cdot Xc) = 2.423 - (2 \cdot 0.0132) = 2.3966$$
$$ZA = 0 \text{ (Remains the same.)}$$

Point B:

Tool radius compensation on the X axis:

$$Xc = 0$$

Tool radius compensation on the Z axis:

$$Zc = r \cdot (1 - \tan \alpha / 2) = 0.032 \cdot (1 - \tan 15) = 0.0229$$

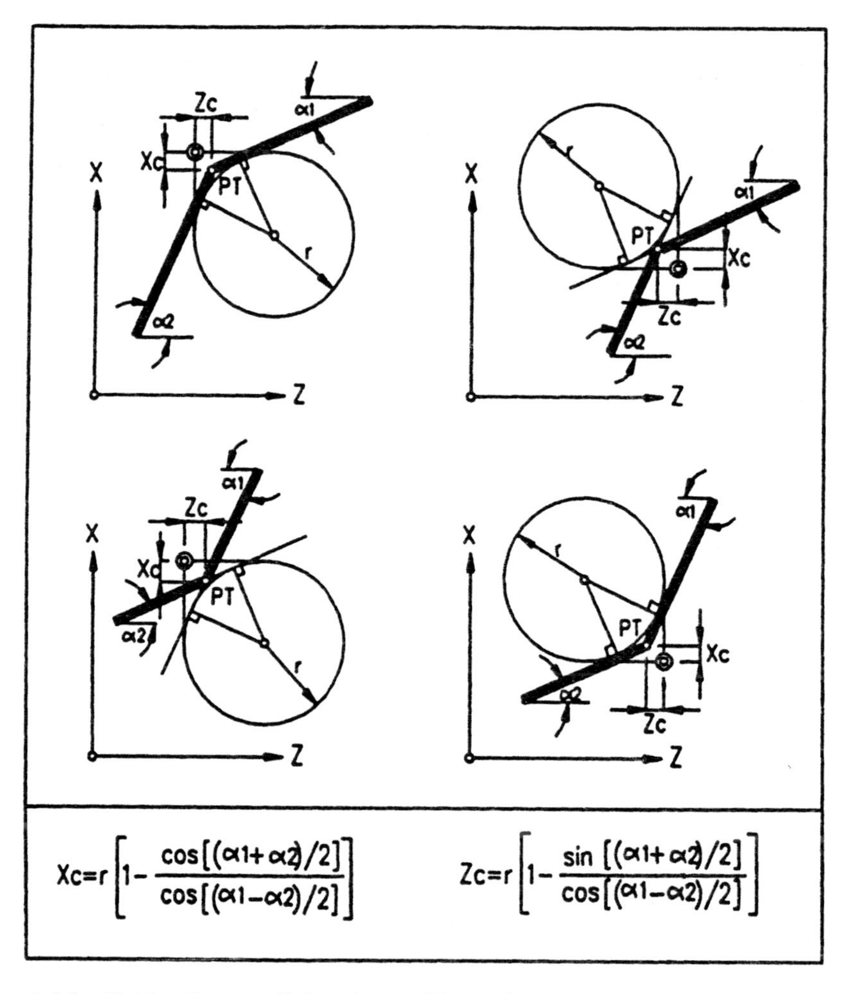

Figure 4–16 Neither line parallel to the machine axis.

Coordinates of point B without compensation:

$$XB = 3.000$$
$$ZB = -0.500$$

Coordinates of point B with compensation:

$$XB = 3.000 \text{ (Remains the same.)}$$
$$ZB = -(0.500 + 0.0229) = -0.5229$$

If tool radius compensation is not entered into the program, the tool will not cut the part properly (Figure 4–18).

At position (a), tool radius compensation is not taken into account at points A and B. The result is a bigger chamfer, since the part is cut along a line p, as programmed, not along line d per the drawing. At position (b), the compensation is

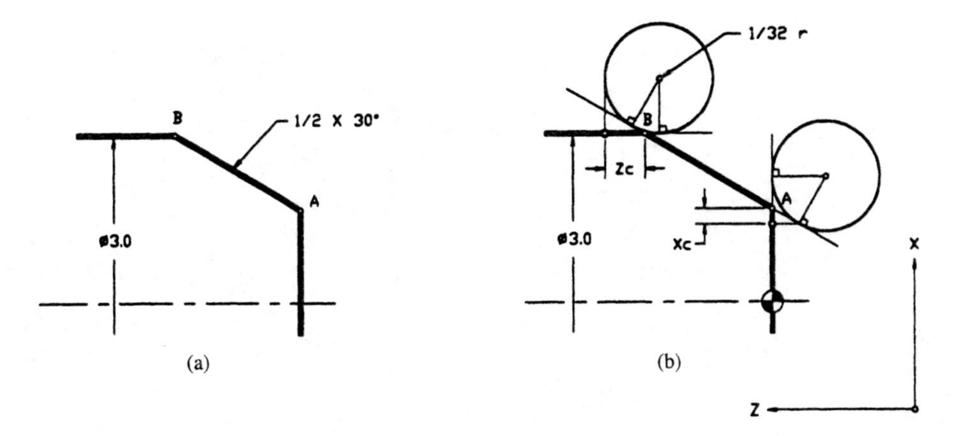

Figure 4–17 Calculating tool radius compensation when cutting parallel to the machine axes.

entered at point A but not at point B. Then the tool starts cutting the part as per the drawing, but finishes away from point B, making a bigger chamfer. At position (c), the compensation is entered at the tool finish point, but not at the tool start point, again resulting in an improperly machined part.

Compensation by Tool Radius Center

Programming by the imaginary tool tip is used only on the lathe, while programming by the tool radius center is used on both the lathe and the machining center. Cutting is performed by a point on the tool periphery, but the tool radius center is

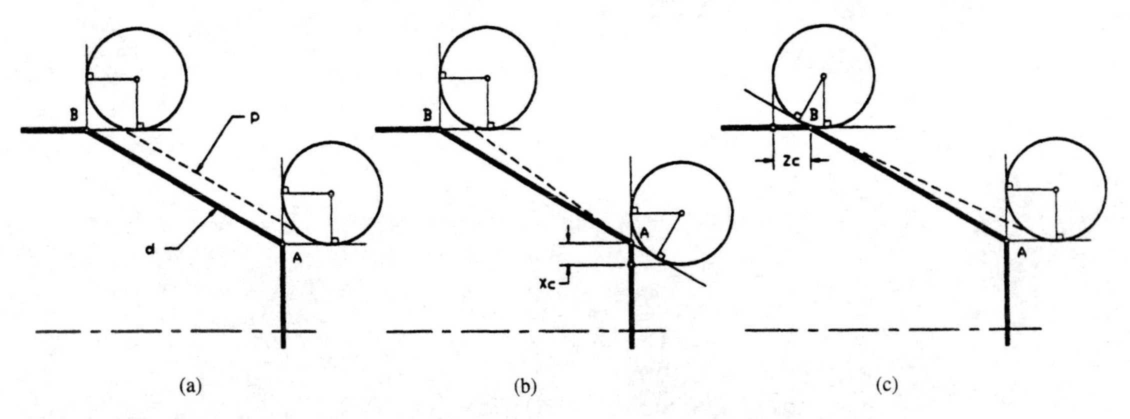

Figure 4–18 When tool radius compensation is not taken into account, the result is unproperly machined part.

used as the tool reference point. This process is referred to as **cutter radius center programming** for the machining center, and as **tool nose center programming** for the lathe.

Figures 4–19, 4–20, and 4–21 all show the intersection of two lines. For each of these situations, there is a formula that can be used to calculate the tool radius compensation when programming either for the lathe or machining center. (When programming for the machining center, Xc and Zc should be substituted for Yc and Xc, respectively.)

In Figure 4–19, the compensation on the lathe Z axis, or machining center X axis, is equal to the tool radius. In general, when the tool is cutting parallel to one axis, the compensation amount for the other axis on the same plane is equal to the tool radius value.

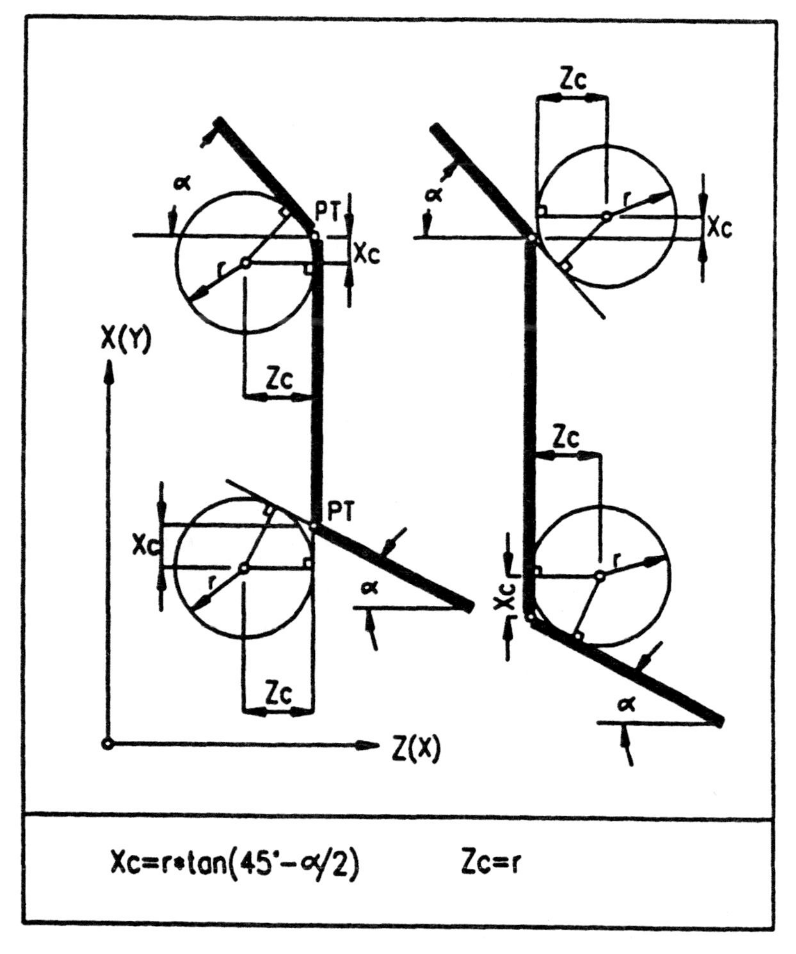

Figure 4–19 One line parallel to the X axis on the lathe, or Y axis on the machining center.

The compensation for the lathe X axis, or the machining center Y axis, is calculated according to the angle and tool radius value. Then it is added or subtracted from the point's coordinate, which depends on the side from which the tool is coming to the intersection.

For the intersection shown in Figure 4–20, the compensation for the lathe X axis, or machining center Y axis, is equal to the tool radius value. On the lathe, the programmer adds a value for the outside cutting that is twice that of the compensation to the point's X coordinate. For the inside cutting, he or she subtracts a value that is twice the compensation value from the point's X coordinate.

The compensation for the lathe Z axis, or machining center X axis, must then be calculated. Then it is added or subtracted from the point's coordinate.

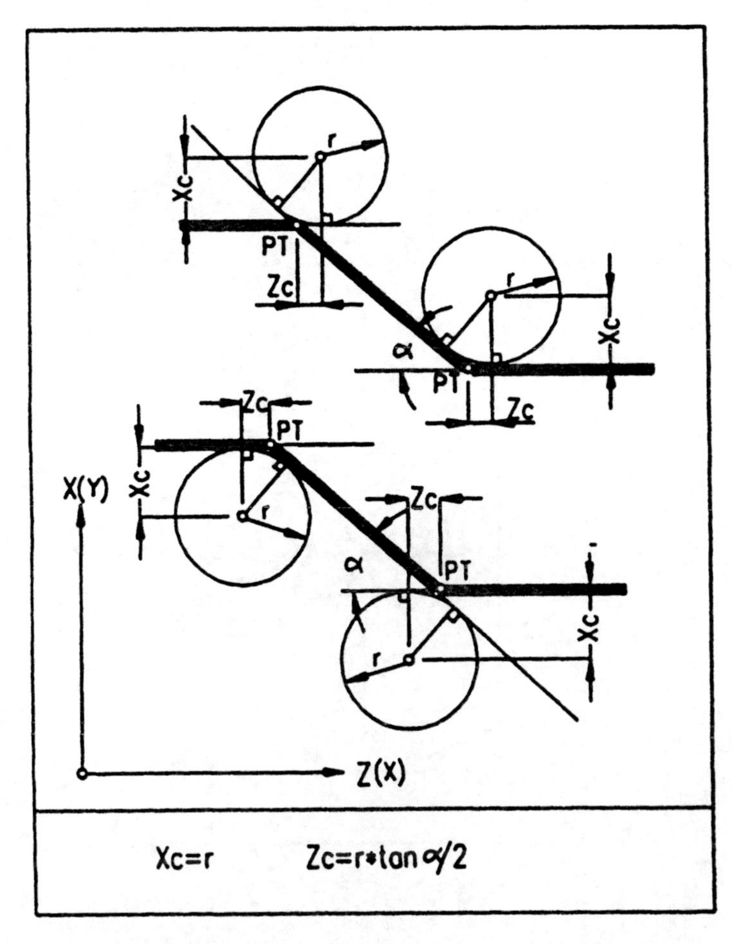

Figure 4–20 One line parallel to the Z axis on the lathe, or the X axis on the machining center.

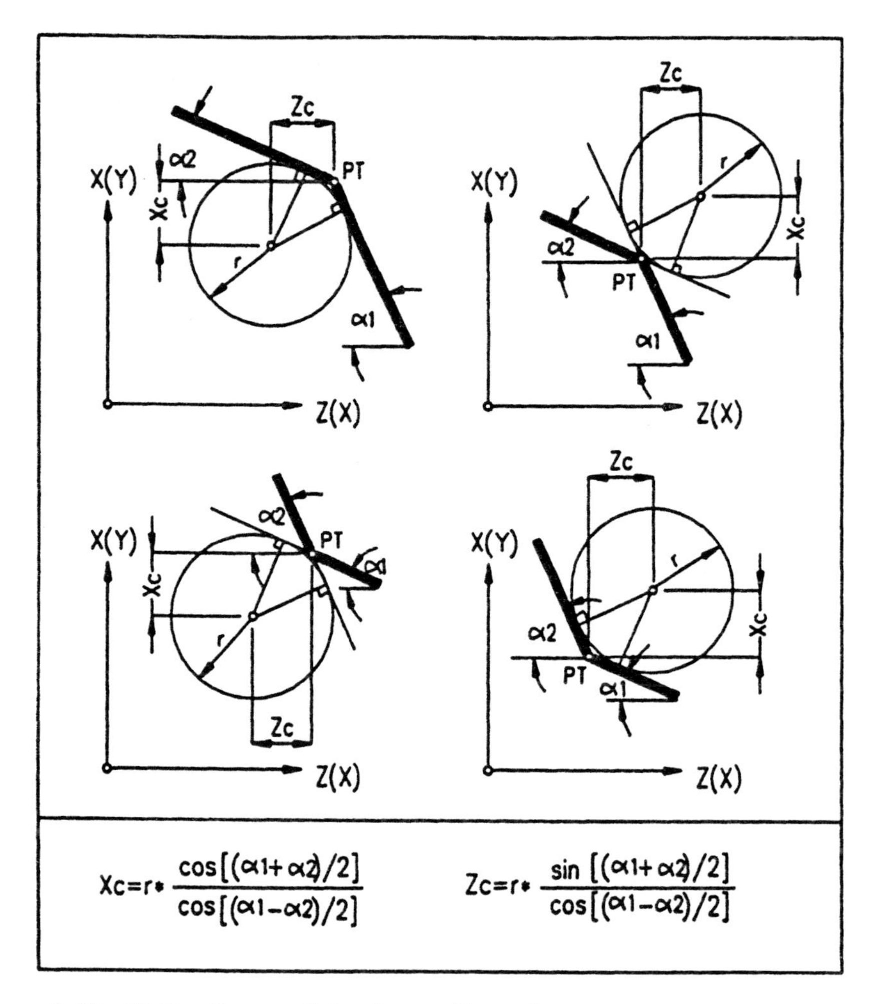

Figure 4–21 Neither line parallel to the machine axis.

When neither line is parallel to the machine axis, as in Figure 4–21, the intersection is made from two angled surfaces. The compensation for either axis is less than the tool nose radius value, which must be calculated.

Following is an example of calculating the coordinates of point A in Figure 4–22 when using a 1-inch-diameter end mill.

At point A, the cutter must be placed perpendicular to both surfaces [position (b)]. The following calculations can be made:

Tool radius compensation:

$$Xc = r \bullet \tan \alpha / 2 = 0.5 \bullet \tan 45 / 2 = 0.207$$

$$Yc = r = 0.5$$

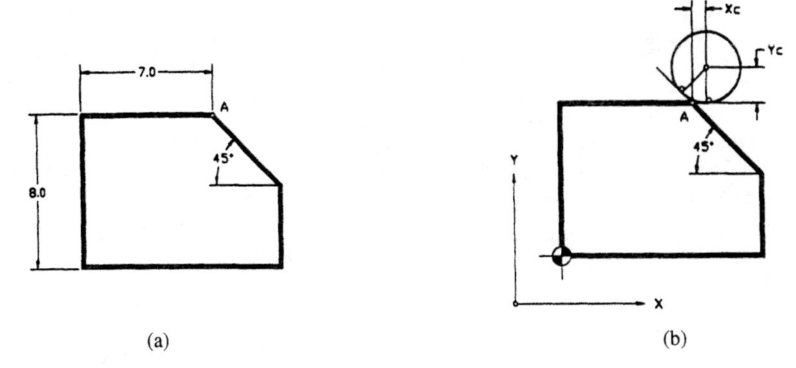

Figure 4–22 Calculating tool radius compensation when programming by tool radius center.

Coordinates of point A with tool radius compensation:

$$XA = 7.0 + Xc = 7.0 + 0.207 = 7.207$$
$$YA = 8.0 + Yc = 8.0 + 0.5 = 8.5$$

When programming by imaginary tool tip, it is easier to read a program since the part diameters and the shoulder lengths are entered without the compensation. Thus they can be programmed directly from the blueprint. For this reason, programming by imaginary tool tip has an advantage over tool nose center programming.

Summary

When programming the points of intersection either by tool nose center or imaginary tool tip, the programmer must take into account compensation amount because of the tool radius. The compensation amount is different for each type of programming because of the different reference points. Particular formulas must be applied for the particular type of programming.

Most of the time, the blueprint does not provide enough detail to program the intersection point directly. Then the coordinates for the intersection point must be calculated. Usually, the programmer draws a sketch of the intersection point using an enlarged view.

Almost any problem in calculating the intersection point can be solved using right triangles. The programmer can draw them in several positions, and then use the triangles to make multiple calculations until the necessary information is obtained. Depending on specific information, the Pythagorean theorem or trigonometric functions may be applied to the right triangles.

Manual tool radius compensation is not much in use on machining centers, except for program editing on the machine. This is still a valid reason why programmers and operators should be proficient in manual programming. Also, a good understanding of this topic will help the programmer use the full power of part programming software.

On lathes, manual tool radius compensation is still in broad use. In many small shops with small work orders, programming by imaginary tool tip is very popular. It allows much greater flexibility than using part programming software.

Key Terms

adjacent side
command point
cosine
cotangent
cutter radius compensation
hypotenuse
intersection point
opposite side
sine
tangent
tool radius compensation

Self-Test

Answers are in Appendix E.

1. The ________________ is a technique of shifting the tool center away from the part in order to perform the cutting by the tool cutting edge.
2. A ________________ is the programmed point on the part.
3. A ________________ is the longest side of a right triangle.
4. The ________________ of a right triangle is the side that lies directly across from the known angle.
5. The ________________ of a right triangle is the side that lies next to the known angle.
6. The most used trigonometric functions are: ________________.

Relating the Concepts

No answers are suggested.

1. Calculate side M from the triangle in Figure 4–23.

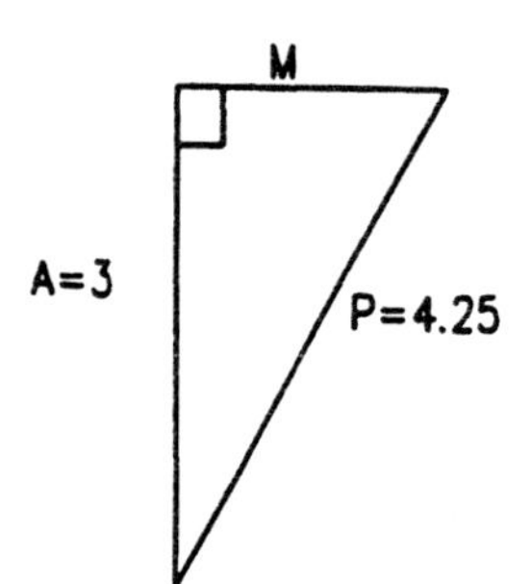

Figure 4–23 Using the Pythagorean theorem.

2. Calculate side T from the triangle in Figure 4–24.

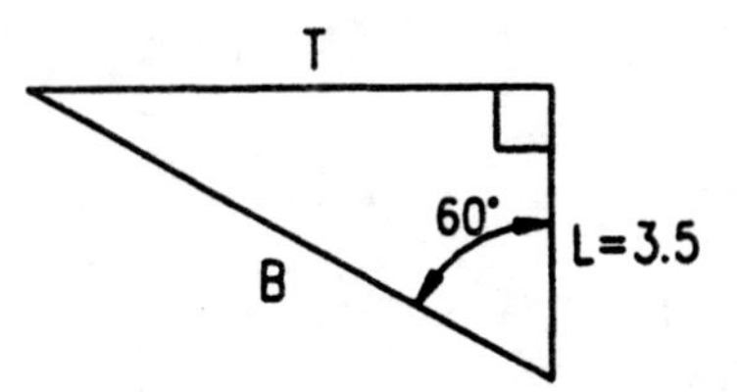

Figure 4–24 Using trigonometric functions.

3. For the shaft in Figure 4–25, calculate the X coordinate of point A and the diameter of the shaft.

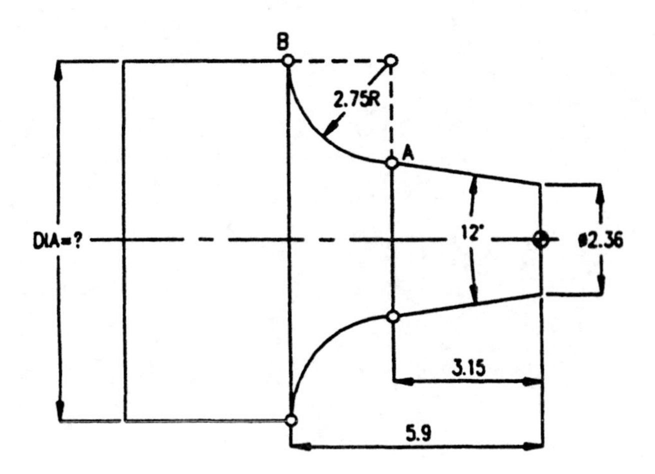

Figure 4–25 Calculating the shaft diameter.

4. In Figure 4–26, using a 0.016 tool nose radius when programming by tool nose center, calculate:
   a. The coordinate of point A
   b. The coordinate of point B

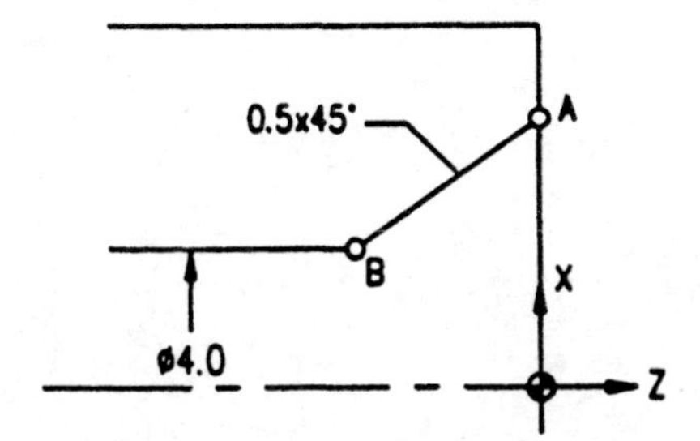

Figure 4–26 Compensation when programming by tool nose center.

5. To cut a particular diameter on the lathe, a block is programmed as 3.678. What part diameter is being cut if a 0.047 tool nose radius is used for programming by tool nose radius center on the lathe when:

   a. OD cutting?
   b. ID cutting?
6. To cut a particular diameter on the lathe, a block is programmed as 3.678. What part diameter is being cut if a 0.032 tool nose radius is used for programming by imaginary tool tip on the lathe when:
   a. OD cutting?
   b. ID cutting?
7. Find the X and Y coordinates for the holes shown in Figure 4–27. Start from the hole that splits the first quadrant and continue in the CW direction.

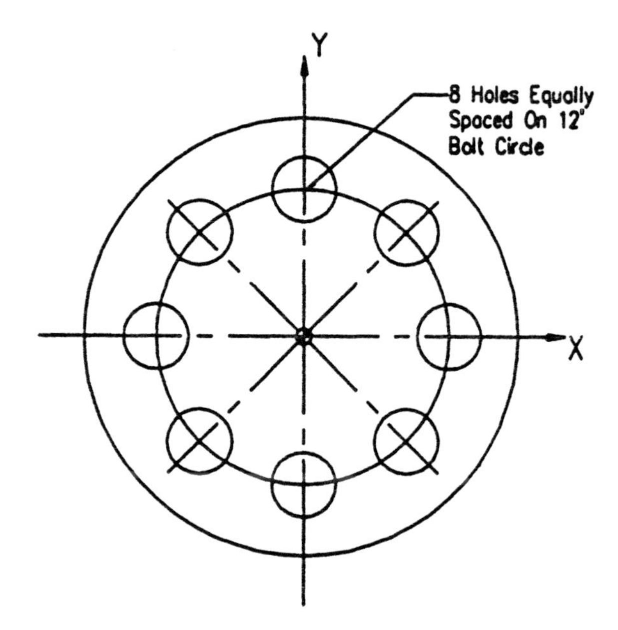

Figure 4-27 Calculating the hole coordinates on the bolt circle.

8. For Figure 4–28, calculate the coordinates of points A and B using a 0.375 radius end mill.

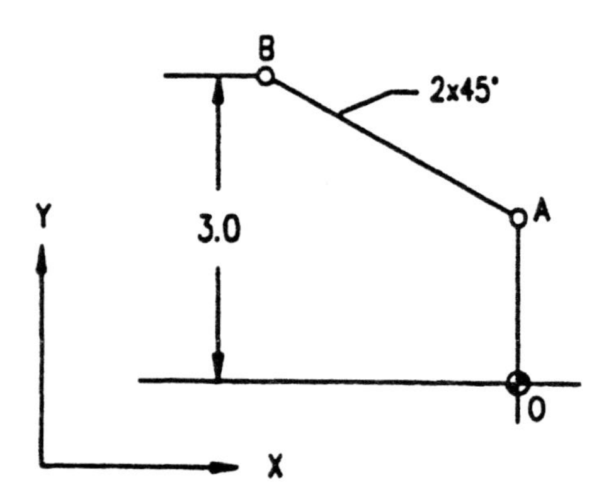

Figure 4–28 Calculating the compensation for a 0.375 radius end mill.

9. For Figure 4–29, calculate the coordinates of points A and B with compensation when programming by imaginary tool tip using a 0.032 tool nose radius.

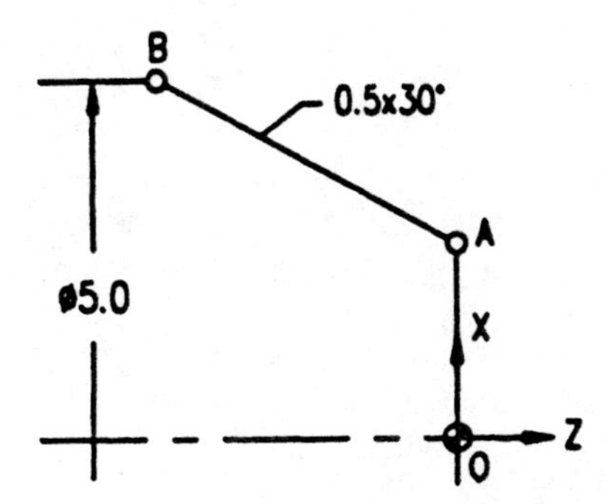

Figure 4–29 Calculating the compensation for a 0.032 tool nose radius.

5 Automatic Tool Radius Compensation

Key Concepts

Programming the Part
Entering the Compensation
Compensation Call/Cancel
A Milling Example
A Turning Example
A Metric Example

Programming the Part

Automatic tool radius compensation is performed by the machine control computer on almost all newer machines, which have powerful built-in software that speeds up programming. When programming by automatic tool radius compensation, the programmer programs the tool to the points of intersection and the integrated software takes care of the compensation. This type of programming is sometimes referred to as *programming the part,* because it is assumed that an imaginary tool with a zero radius is used when writing the program. Literally speaking, we program the points on the part, not the tool reference point, as we do when programming by the manual tool radius compensation method.

Using this method when milling or turning, the tool automatically shifts away from the part. The shift amount depends on tool values stored in the geometry offset using manual data input (MDI). Geometry offsets are used in programming by automatic tool radius compensation for the following purposes: using a different tool diameter than was originally programmed; varying the part size while machining; making a series of cuts when roughing and finishing using the same programmed data; and adjusting for tool wear.

In automatic tool radius compensation, it is possible to use a different tool diameter instead of the one originally programmed. For example, let's say that the program calls for a 1 1/16-inch-diameter cutter, but this size is not available. The closest size cutter available is 1.0 inch in diameter. The operator enters a 0.5 inch radius value into the particular geometry offset. The control shifts the tool away from the part for this amount, and the part will be machined as if the original cutter size was used. Note that no change in the program is needed, since the control recalculates the tool path according to the value in the geometry offset.

When machining using automatic tool radius compensation, it is easy for the operator to vary the part size. For instance, if the part is machined oversize, he or she decreases the geometry offset value by the appropriate amount. Say the part is 0.005 inch oversize; the operator should change the R value entered from 0.500 to 0.495. This will shift the tool 0.005 inch closer to the part, producing the desired size. If the part is machined under size, the operator should increase the R value, making the control think that a cutter of larger size is being used, even if it is not. The increased R value will move the cutter centerline farther away from the part for an appropriate amount, producing a part that is cut less than before.

Automatic tool radius compensation makes it possible to make a series of cuts when roughing and finishing using the same programmed data. For example, a 0.510 inch radius may be entered for offset 031, and 0.500 inch radius may be entered for offset 032. When using offset 031, the tool leaves a 0.010 inch for finishing. The part will be cut to finish size using offset 032. Thus, the programmed data is the same; only the offset value is different.

Entering the Compensation

To enable the control to use tool radius compensation, the programmer must call the particular geometry offset in the program when initializing the compensation, and the operator must enter the size of the tool radius into the particular geometry offset using MDI. On the machining center, the operator enters the cutter radius value into the tool offset register using the D address. The machine is now ready to use the tool radius compensation. The same address is used to call the particular geometry offset in the program. This means that the geometry offset number can be labeled differently from the tool length offset number. For example, H7 may be the tool length offset for one particular tool, and D32 may be the geometry offset for that tool.

On the lathe, preparing the machine for the tool radius compensation is a bit different, because there are eight standard tool nose vectors that can be used (Figure 5–1).

A **tool nose vector** is defined by its direction and size. The direction is shown as an arrow from the center of the tool nose radius to the intersection of the two cutting edges of the tool nose. The size of the tool nose vector is equal to the size of the tool nose radius.

In practice, the tools used most on the lathe are the tools for ordinary turning and boring. Consequently, the tool nose vectors used most are number 3 for turning and number 2 for boring. Note that these tool nose vectors are opposite for machines with left-hand coordinate systems.

On the lathe, the R address is often used to enter the size of the tool nose radius into the geometry offset, although on some lathes the D address may be used for this purpose. Also, the tool nose vector number must be entered into the particular geometry offset.

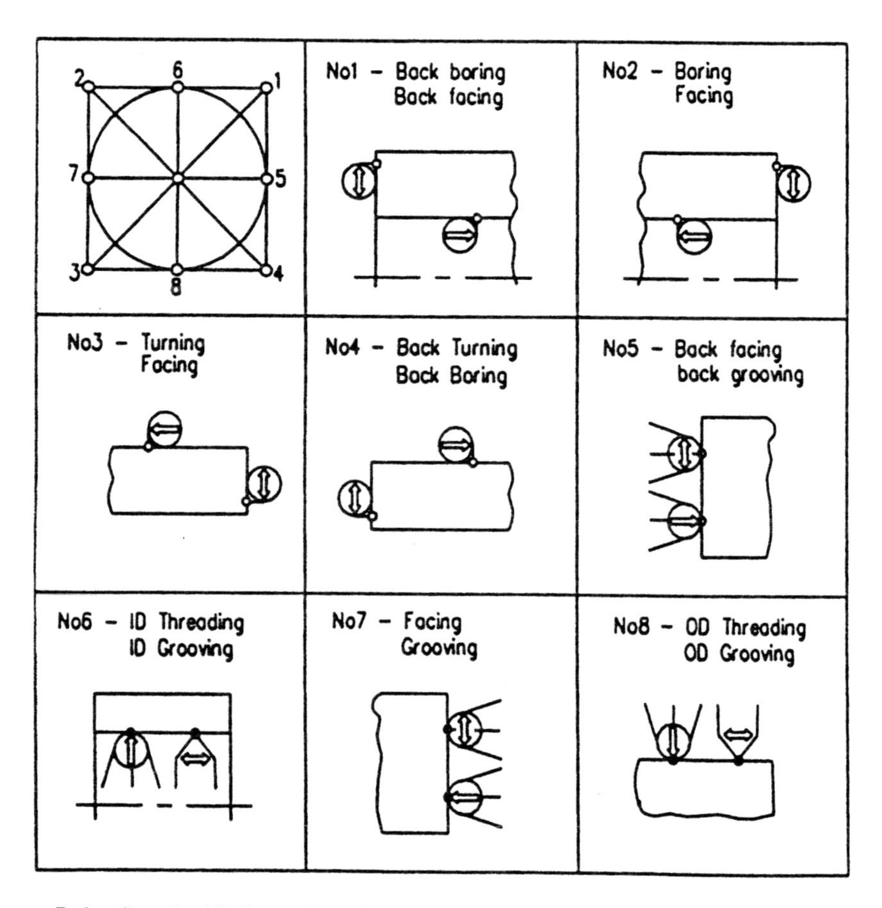

Figure 5–1 Standard lathe tools and tool nose vectors.

To call the particular geometry offset in the lathe program, the programmer uses the R or D address. The geometry offset can then be labeled differently from the tool length offset. However, on some lathes the compensation can be initialized without specifying any particular address. The control uses the same number for the geometry and tool length offsets. For example, if tool length offset number 03 is used, the control will use the same number for the geometry offset.

Compensation Call/Cancel

Once given the appropriate compensation code, the tool will change its position to the compensated position on the next programmed move. It continues cutting in compensation mode until compensation is cancelled. Then the tool position changes from **compensated position** to **uncompensated position** on the next programmed move.

Before returning the tool to home position, the programmer must cancel compensation. If he or she does not do so, the control will remember the compensation amount and use it later in the program. This may cause the following problems: compensation may be used for a tool that does not need it, such as a drill; the compensation amount may build on the compensation amount for another tool; or an alarm may be raised on some machines.

Compensation is initialized or cancelled in the program when the control reads the following codes:

G41 **Compensation left.** The G41 code moves the tool left of the part surface by the amount entered into the geometry offset. The left side of the part is determined by looking from behind the tool in the direction the tool is moving. On the lathe, the G41 code is normally used for facing and inside diameter cutting. On mills this code is used for climb milling.

G42 **Compensation right.** The G42 code moves the tool right of the part surface by the amount entered into the geometry offset. The right side of the part is determined by looking from behind the tool in the direction the tool is moving. On the lathe, the G42 code is normally used for outside diameter cutting. On mills this code is used for conventional milling.

G40 **Compensation cancel.** The G40 code cancels the compensation initialized by the G41 or G42 code. The control returns to uncompensated mode, and the tool then moves directly on the programmed shape.

The compensation codes have the same purpose and are programmed in the same way on both the lathe and the machining center.

Tool radius compensation may be called and cancelled for one particular tool as many times as desired. On the lathe, different tool nose vectors cannot be used for the same tool, but different sizes of the tool nose radius may be used if desired.

Sometimes it is difficult to distinguish whether the G41 or G42 code is to be used. Figure 5–2 shows the application of these codes.

At position (a), the direction of tool travel is indicated by the index finger (I), while the side on which to offset the tool is indicated by the thumb (T). When

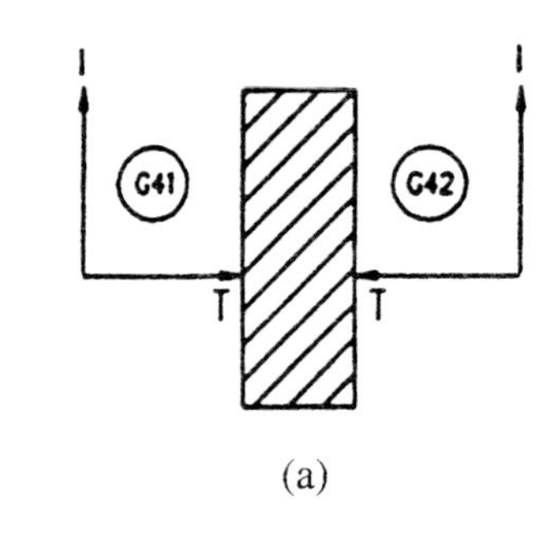

(a)

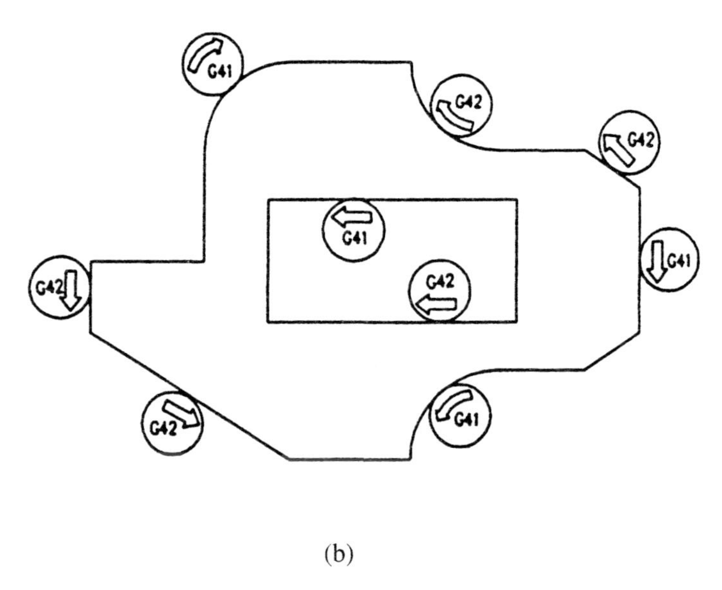

(b)

Figure 5–2 Application of the G41 and G42 codes.

using the left hand, compensation is needed at the left of the part surface; thus G41 should be programmed. When using the right hand, compensation is needed at the right of the part surface; then G42 should be programmed. Some application examples of using G41 and G42 are also shown at position (b).

When executing the program, the control reads several lines at once in order to position the tool for the next move. This is important when automatic tool radius compensation is used, since the tool radius must always be set perpendicular to both surfaces, present and upcoming, as illustrated in Figure 5–3.

Before moving the tool from point A, the control reads the next line in which the compensation is initialized by the G41 code. Then it knows that the tool is to be placed perpendicular to the next line, which is facing along the X axis. Consequently, the tool moves to the coordinates of point B and, at the same time, the tool nose center is placed perpendicular to the part face. This continues for each block in the program. This way of reading the program in blocks helps the control not

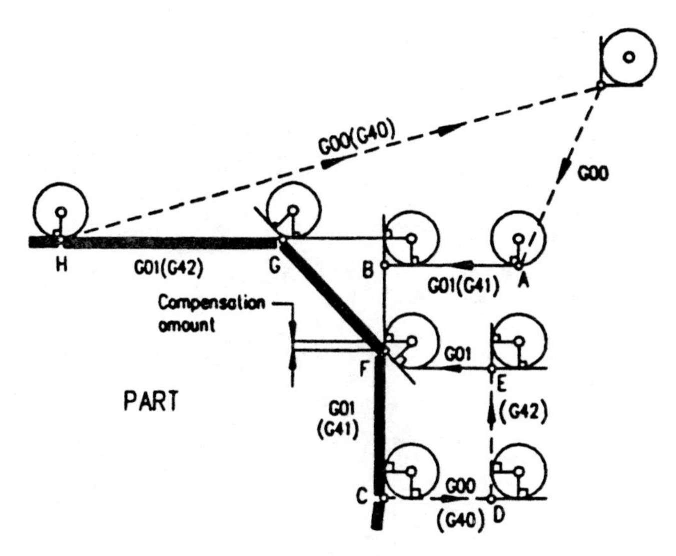

Figure 5–3 The tool is set perpendicular to the line of the next move.

only to execute the instructions, but also to find erroneous ones in order to prevent their execution and generate an alarm to warn the operator of the error.

When cancelling compensation, the tool should have room to retract; otherwise, overcutting will occur. If there is no room for the tool to retract, it should be moved a safe distance before cancelling compensation. Afterwards, compensation may be cancelled at any convenient point. This is usually done while the tool is returning to the home position.

Sometimes compensation has to be cancelled immediately after cutting a chamfer, taper, or slope. Say the cutting has to be stopped at PT1. The compensation is still on and the tool nose center is a distance away from PT1. The amount of compensation is Zc [Figure 5–4, position (a)].

At position (a), the tool may be programmed to move away from the part surface on the X axis. In the same block, the compensation may be cancelled, as below:

G00 G40 U-0.1; (Compensation cancel by G40. Retract in rapid by G00, programmed as an incremental move on the X axis.)

When the G40 code reads in, the tool nose center is positioned perpendicular to the last line of motion, which is the line parallel to the Z axis. The tool changes its position from the compensated to the uncompensated position. As a result, the tool nose center is positioned in the same line with PT1, because the compensation on the Z axis is equal to zero. Consequently, overcutting occurs as the result of cancelling the compensation. As seen from position (b), the overcut is made on the chamfered surface. To prevent this from happening in a move-away block, a chamfer value has to be programmed with I and K, as follows:

G40 U-0.1 I-0.25 K-0.25;

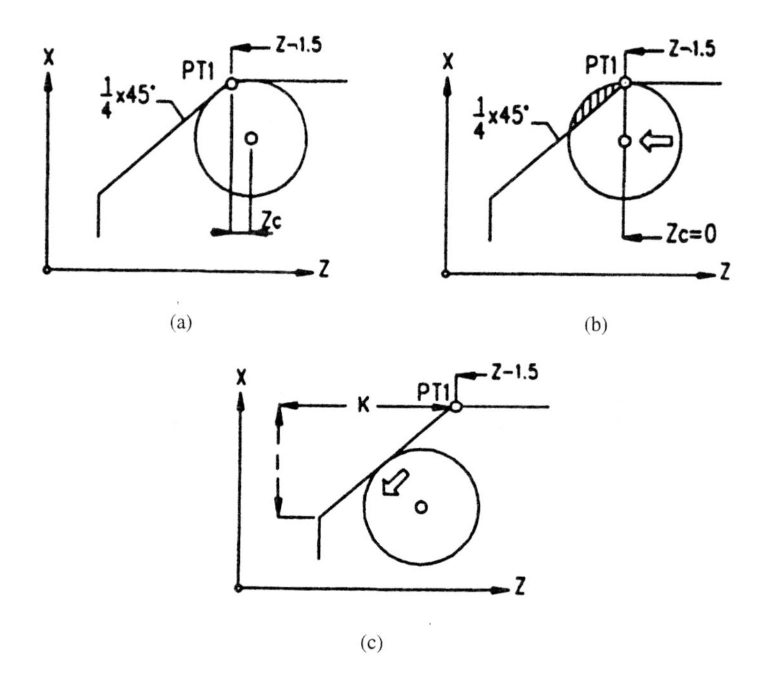

Figure 5–4 Cancelling the compensation to prevent the overcut.

I and K move the tool at an angle when cutting a chamfer or taper. The move on the X axis is expressed by the I value, while the K address controls the move on the Z axis. The control then cancels the compensation by placing the tool nose center perpendicular to the chamfer. Thus this instruction cancels the compensation at PT1 and moves the tool away from the part without hurting the chamfered surface. This is illustrated at position (c).

In ordinary programming practice, it is almost always possible to position the tool at a safe distance before the compensation is cancelled. For example, to prevent overcutting, we can simply move the tool out from the bore and then cancel the compensation. This is illustrated in the following series of instructions:

```
G01 Z-1.5; (Cutting up to PT1.)
G00 U-0.1; (Moving tool safely away.)
G40 Z0.3; (Compensation cancel, tool out from the bore.)
```

On most controls, when calling and cancelling tool radius compensation, the move distance must be more than the tool radius value. When programming for the lathe, the move distance should be at least twice the tool nose radius value on the X axis and one radius value on the Z axis. There are several alarm messages that may result if this simple rule is not followed.

When programming using the automatic tool radius compensation, always keep in mind that compensation, once initialized, will stay in effect until cancelled. This means that the next tool will use compensation even if it is not programmed. If this tool also calls the compensation, it will build up on the previous one. If a tool is programmed without compensation (such as a drill or tap), it may behave in a strange way, or break, because it is not positioned on centerline. Thus, always make sure that the compensation is cancelled. To make sure that the compensation is cancelled, some programmers enter the G40 code in a **safety line** at the beginning of the program:

N1 G00 G40 G80;

In this line, G40 cancels compensation if programmed before, while G80 cancels any canned cycle previously programmed. If there is no previously programmed compensation or canned cycle, this line will not affect the program. Thus it does no harm to program it if desired, and it may help on some occasions.

The tool radius compensation called by the G41 or G42 code is engaged using either the G00 or G01 code on the same line. Usually, the compensation is called when the tool is approaching the part. Then the particular compensation code is called when feeding the tool into the part. For example:

G01 G41 Z0 F0.015; (Feeding to a Z0 on the lathe. Compensation call by G41.)

A radius smaller than that of the tool nose radius cannot be programmed using tool radius compensation. Instead, a sharp corner should be programmed using the natural radius of the tool. Following are examples of applying the automatic tool radius compensation when programming for the machining center or the lathe, either in inches or in metric units.

A Milling Example

The machining center example involves writing the program for the part illustrated in Figure 5–5. Notice that there are no holes on the part for the bolts to clamp it on the table. Hence the part must be machined in two operations: While the part is being machined on its left-hand side, the right-hand side of the part is clamped, and vice versa. When the reclamping operation takes place, the part does not have to be moved, and the same reference point can be used as the part origin. If the job involves multiple parts, they can all be machined on one side, then on the other.

For this example, the part is 0.75 inch thick, and it is assumed that the rough profile is already cut. Thus, only the finishing pass should be programmed. The material is mild steel and a 1.0-inch-diameter end mill will be used. The operator should enter the size of the tool radius, which should have been written on the setup sheet, into the particular geometry offset using MDI.

G0 G40 G90 G54 G43 X-0.6 Y-0.6 Z2.0 H1 S750 M03; (G40 cancels compensation if previously programmed. G54 selects the coordinate system. X, Y, and Z position the tool. G43 and H1 call the tool length offset. Spindle start by M03 using 750 RPM.)

Z-0.8; (Lower the tool on the Z axis.)

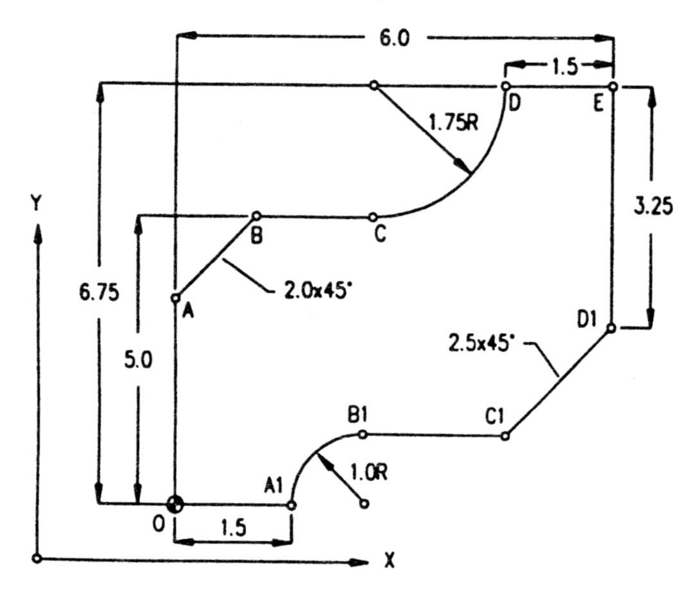

Figure 5–5 Programming automatic tool radius compensation for machining center.

G01 G41 D31 X0 F8.75 M08; (Cutter compensation left by G41. Geometry offset called by D31. This will place the cutter centerline 0.5 inch from the part surface on the X axis, although X0 is programmed.)
Y3.0 F3.75; (Point A.)
G91 X2.0 Y2.0 F3.0; (Point B in incremental programming.)
G90 X2.75; (Point C in absolute programming.)
G03 X4.5 Y6.75 R1.75; (Point D, arc cutting.)
G91 G01 X2.1; (Straight cutting in incremental programming to pass point E for at least the size of the tool radius.)
G0 G90 Z8.0 M09; (Retract on the Z axis. Coolant stop.)
G40 (Tool radius compensation cancel.)
G0 Y12.0; (This instruction will move the table out for the manual clamp change.)
M00; (Program stop—reclamp the part.)

S750 M03; (Spindle start.)
G00 G90 X-0.6 Y-0.6; (Rapid to position to cut the other side.)
Z-0.8; (Lower the tool on the Z axis.)
G01 G42 D31 Y0 F3.75 M08; (Cutter compensation right by G42. Geometry offset called by D31. This will place the cutter centerline 0.5 inch from the part surface in the Y axis, although Y0 is programmed.)
X1.5; (Point A1.)
G02 X2.5 Y1.0 R1.0; (Point B1, arc cutting.)
G01 X3.5; (Point C1.)
G91 X2.5 Y2.5; (Point D1 in incremental programming.)
G90 Y7.3; (Straight cutting to pass point E. The programmer must program the tool to move past the point by at least cutter radius value. Otherwise, this surface will not be machined completely.)

G0 Z2.0 M09; (Retract on the Z axis. Coolant stop.)
G40 (Tool radius compensation cancel.)
G28 Y0 Z0; (Return home on the Y and Z axes. The Y instruction will move the table out to help the operator in manual operations.)
M30; (Program stop.)

The tool length offset H1 was not cancelled after the first operation; therefore, it does not have to be called again for the second operation.

Different tool geometry offsets may be used when roughing and finishing passes are needed. For example, when roughing the programmer can assign D30 = 1.010, and for finishing D31 = 1.000. Then, he or she can call the program twice or use a subprogram. In either case, the roughing tool will leave 0.010 inch for finishing.

A Turning Example

This example, which illustrates the use of automatic tool radius compensation on the lathe, uses the part shown in Figure 5–6. The part is chucked from the outside diameter, allowing the bore to be machined all the way through up to point E1. The outside diameter is turned up to point E. Only the finish profile will be cut. The material is mild steel.

N1 G50 X15.0 Z3.0 S1000 M42; (Coordinate system preset for turning tool speed limit. Low range.)
G00 T0100; (Indexing the tool.)
G96 S530 M03; (Spindle start.)
G00 X4.8071 Z0.1 T0101 M08; (Rapid to point A, which is calculated as $5.5 - 2 \cdot (0.6 \cdot \tan 30) = 4.8071$).

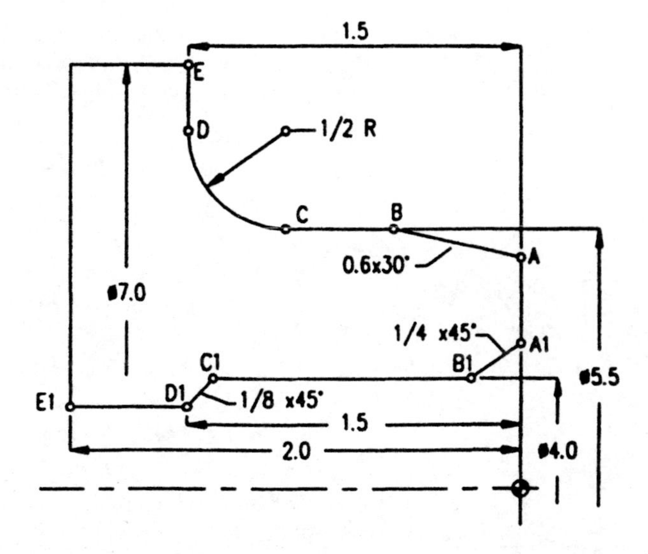

Figure 5–6 Programming automatic tool radius compensation on the lathe.

G01 G42 Z0 F0.015; (Feeding to Z0. Compensation right by G42.)
X5.5 Z-0.6 F0.009; (Point B, chamfer cutting.)
Z-1.0 F0.013; (Point C, calculated as 1.5 – 0.5 = 1.0.)
G02 X6.5 Z-1.5 R0.5 F0.009; (Point D, radius cutting.)
G01 X7.1 F0.011; (Cutting along the X axis to pass point E.)
G00 G40 X15.0 Z3.0 T100 M9; (Compensation cancel. Return home. Offset cancel.)
M01; (Optional program stop.)

N3 G50 X15.0 Z3.0 S900 M42; (Coordinate system preset for boring bar. Speed limit. High range.)
G00 T0300; (Indexing the tool.)
G96 S530 M3; (Spindle start.)
G00 X4.5 Z0.5 T0303 M08; (Point A1, calculated as: 4.0 + 2 • (0.25 • tan 45) = 4.5).
Z0.1; (Closer to the part on the Z axis.)
G01 G41 Z0 F0.015; (Feeding to Z0. Compensation left by G41.)
X4.0 Z-0.25 F0.009; (Point B1.)
Z-1.375 F0.012; (Point C1.)
U-0.25 W-0.125 F0.01; (Point D1 in incremental programming.)
Z-2.05 F0.013; (Finishing the bore length passing point E1.)
U-0.1; (Tool is moved away from the part surface in incremental programming.)
G00 Z1.0; (Tool is pulled out of the bore.)
G40 X15.0 Z3.0 T0 M9; (Compensation cancel. Tool is returned to the home position.)
M05; (Spindle stop.)
M01; (Optional program stop.)
M30; (Program stop and rewind.)

Prior to running the following program, the operator should enter the size of the tool nose radius into the particular tool offset registers. He or she should also enter the tool radius vectors: number 3 for turning and number 2 for boring.

If facing is to be performed, it is not necessary to use tool radius compensation; if it is desired, the programmer can adjust the program:

G50 X15.0 Z3.0 S1000 M42; (Coordinate system preset.)
G00 T0100; (Indexing the turning tool.)
G96 S530 M03; (Spindle start.)
G00 G41 X5.6 Z0 T0101 M08; (Rapid to position for facing. Offset call. Compensation left by G41.)
G01 X3.9 F0.01; (Facing. In order to clean up the part face completely, the tool must be programmed to move below bore diameter for at least twice tool nose radius value.)
G00 G40 Z0.1; (Compensation cancel by G40.)
X4.8071 Z0.05; (Positioning for chamfer cutting.)
G01 G42 Z0 F0.01; (Feeding to Z0. Compensation right by G42.)
X5.5 Z-0.6 F0.009; (Point B, chamfer cutting.)
...; (Continuing.)

Note that when switching from one compensation mode to the other, the programmer must program the compensation cancel code in between.

A Metric Example

When programming with automatic tool radius compensation in metric, the size of the tool nose radius must be entered in millimeters to ensure uniformity of dimensions. Following is the programming example for the part illustrated in Figure 5–7. The material is 4140 steel. Again, we are programming only a finish profile.

G21; (Programming in metric.)
G50 X300.0 Z75.0 S1300 M42; (Coordinate system preset. Speed limit. High range.)
G00 T0100; (Indexing the tool.)
G96 S500 M03; (Spindle start.)
G00 X90.0 Z3.0 T0101; (Rapid to position. Tool offset call.)
G42 G01 Z0 F0.35; (Feeding to Z0. Compensation right by G42.)
X100.0 Z-5.0 F0.23; (Chamfer cutting.)
Z-40.0 F0.3; (Straight cutting.)
G02 X140.0 Z-50.0 R10.0 F0.2; (Radius cutting.)
G01 W-30.0 F0.3; (Incremental move up to the taper start point.)
X160.0 W-80.0 F0.25; (Taper cutting in incremental programming.)
W-30.0 F0.3; (Straight cutting in incremental programming.)
X205.0; (Shoulder cutting.)
G00 G40 X300.0 Z75.0 T100 M09; (Return home. Compensation cancel. Offset cancel.)
M1; (Optional program stop.)

If the standard 0.8 mm tool nose radius is to be used, when calling and cancelling the compensation, the distance from the tool cutting edge to the part surface is more than enough.

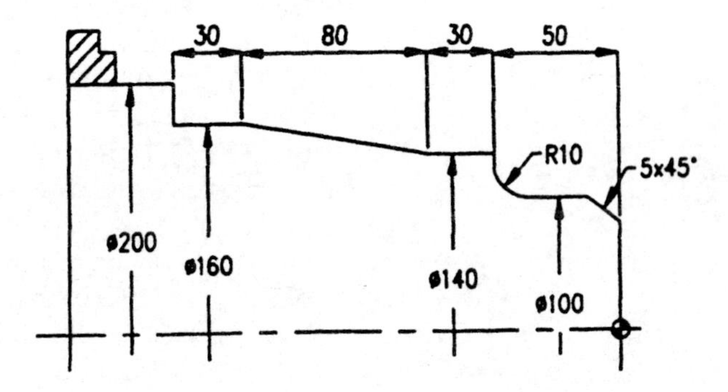

Figure 5–7 Programming automatic tool radius compensation using metric units.

Summary

When using automatic tool radius compensation, the programmer actually programs the part points of intersection, and the machine software does all the necessary calculations in order to find the compensation values. It makes programming, as well as program editing on the machine, easier.

When milling or turning, the tool is automatically shifted away from the part. The shift amount depends on the value stored into the geometry offset using manual data input (MDI).

Geometry offsets may be used for the following purposes: to compensate for a different tool diameter instead of one that was originally programmed; to vary the part size during machining; to make a series of cuts when roughing and finishing using the same programmed data; and to adjust for tool wear.

On the machining center, the geometry offset is usually assigned by the tool radius value. On the lathe, the tool nose radius value is entered with the tool nose radius vector.

Automatic tool radius compensation is initialized by either the G41 or G42 code. The G41 code moves the tool left of the part surface by the amount entered into the geometry offset. On lathes it is used for facing and inside diameter cutting; on mills this code is used for climb milling. The G42 code moves the tool right of the part. In lathe programming it is used for the outside diameter cutting; on mills it is used for conventional milling. The G40 code cancels the compensation initialized by the G41 or G42 code.

Key Terms

automatic tool radius compensation
compensated position
compensation cancel
compensation left
compensation right
safety line
tool nose vector
uncompensated position

Self-Test

Answers are in Appendix E.

1. When programming by _______________, the tool is programmed to the points of intersection and the integrated software takes care of the compensation.
2. A _______________ is defined by its direction and size.
3. Once given the appropriate compensation code, the tool will change its position to _______________.

4. When compensation is cancelled, the tool position is changed from compensated to ______________.
5. ______________ is programmed by the G41 code.
6. ______________ is programmed by the G42 code.
7. ______________ is programmed by the G40 code.
8. The line with the G40 and G80 codes at the beginning of the program is called the ______________.

Relating the Concepts

No answers are suggested.

1. Explain how to program contour milling for finishing and roughing passes.
2. The programmed cutter diameter is 1.0 inch. The closest available size is 1.125 inch. What would you do to finish the job?
3. When using automatic tool radius compensation, why is it important for the control to read several lines ahead?
4. For the part shown in Figure 5–8, write a program in metric using automatic tool radius compensation. Use a 0.8 mm tool nose radius.

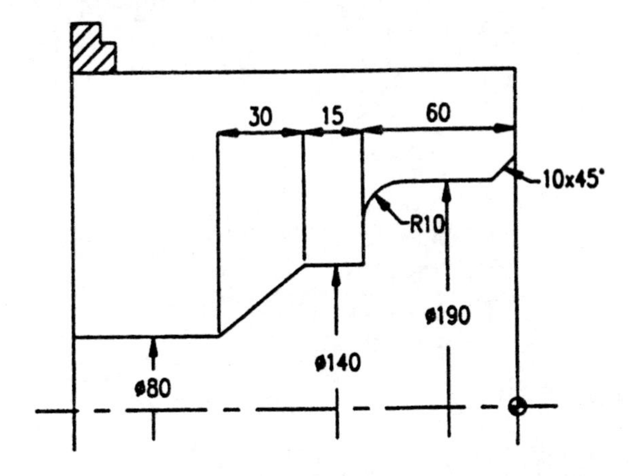

Figure 5–8 Programming the boring operation in metric using the automatic tool radius compensation.

5. When calling and cancelling compensation on the lathe, the tool move must be at least twice the radius value for the X axis and one radius value for the Y axis. Explain.

6 Tooling Features

Key Concepts

Types of Cutting Tools
Cutting Tool Materials
Carbide Inserts Classification
Cutting Fluid
Operating Conditions and Tool Life
Adaptive Control

Types of Cutting Tools

CNC machine tools are made to improve productivity and reduce manufacturing costs. The programmer and operator must have a thorough knowledge of cutting tools and their proper use; selecting the right cutting tools is essential to obtaining the best results. There are two major types of cutting tools used on CNC machines, **single-point tools** and **multipoint tools.** Single-point tools are used primarily in turning and boring operations on turning centers, but they are also used for boring on machining centers. In modern metal-cutting operations, all single-point cutting tools use **indexable** carbide inserts, meaning more than one cutting edge can be used before the insert is no longer usable. They are available in a variety of shapes (Figure 6–1).

Figure 6–1 Single-point tools for turning and boring operations. (Courtesy Valenite, Inc.)

Multipoint tools, such as milling cutters, end mills, shell end mills, and reamers are used for milling, drilling, reaming, counterboring, and countersinking operations. Smaller sizes of multipoint cutting tools have two or four cutting edges, while larger sizes have more than four (Figure 6–2).

As shown in Figure 6–3, the number of cutting edges increases with the cutting tool diameter. This is because the larger tool must be programmed to rotate

Figure 6–2 Multipoint cutting tools. (Courtesy Valenite, Inc.)

Figure 6–3 On multipoint cutting tools, the number of cutting edges increases with the cutting tool diameter. (Courtesy Sandvik Coromant Company.)

more slowly. To achieve the same surface finishing as smaller tools, these tools are designed with more cutting edges.

Cutting Tool Materials

The cutting tool must be harder than the material being cut, and over the years, harder and tougher part materials have necessitated harder and tougher cutting tool materials. These materials allow cutting at a higher production rate. The following materials are used most often for cutting tools:

1. High-speed steel (HSS)
2. Cemented carbides
3. Ceramics
4. Diamond
5. Cubic Boron Nitride (CBN)

High-speed steel (HSS) cutting tool materials are alloy steels, which contain primarily tungsten and cromium, with a small percentage of cobalt, vanadium, and molybdenum. HSS tools are mainly end mills and drills. These tools are used when cutting with lower cutting speeds (up to 150 SFM in 4140 steel) and lower tool contact temperatures (up to 550°C or 1022°F). HSS tools can be resharpened easily, which is their main advantage.

Cemented carbides, a form of carbide (a mix of carbon and a metal), are composed of tungsten carbide, tantalum carbide, or titanium carbide, with cobalt as a bond. They are made by **sintering,** a process of melting and pressing the individual particles together. A variety of cemented carbide tools and inserts are used in modern manufacturing, either on machining or turning centers (Figures 6–4 and 6–5).

Cemented carbide tools are used when cutting with higher cutting speeds (up to 400 SFM in 4140 steel) and higher tool contact temperatures (up to 1200°C or 2192°F). In general, these tools cannot be resharpened. The use of coatings such as aluminum oxide inhibit heat buildup, resulting in longer tool life. Because of higher cutting speeds and longer tool life, the **coated carbides** are widely accepted as a cutting tool material for milling and turning applications.

Ceramics are cemented oxide tool materials made by sintering in a similar fashion as cemented carbides. However, they are harder than carbides and allow cutting at higher temperatures and cutting speeds. Ceramic cutting tools are

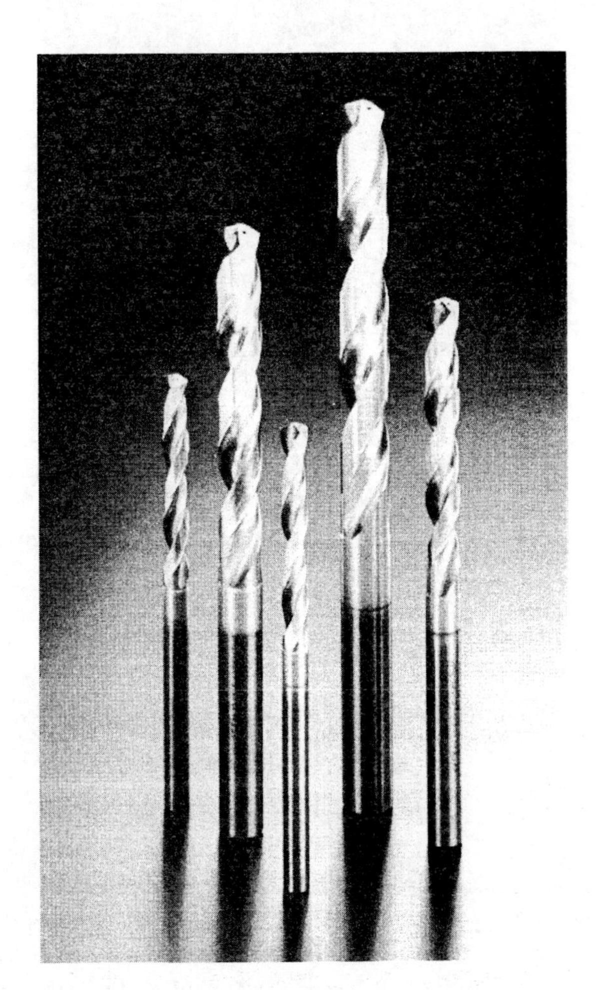

Figure 6–4 Solid carbide, regrindable drills with coolant holes. Coolant holes in flutes assist in chip disposal, allowing deeper drilling depth. (Courtesy Mitsubishi Materials U.S.A. Corp.)

Figure 6–5 A variety of carbide inserts. The multifunction chipbreakers allow for lower cutting forces, smooth chip evacuation, and superb chip control in a variety of turning operations. The inserts are coded and marked for easy identification and use. (Courtesy Sandvik Coromant Company.)

available only in insert form since they are very brittle. Ceramic tools are used for machining hard materials with high speeds using a small depth of cut (sometimes just 0.001 inch). As with carbide inserts, ceramic tools cannot be resharpened.

Diamond is the hardest material known. Synthetic diamond, known as **polychrystaline-diamond** or **PCD,** is more suitable for cutting tools than natural diamond. Polychrystaline tooling is often used in precision machining and finish operations of nonferrous (nonmetallic) materials. It is extremely important that polychrystaline-diamond cutting tools are used only with compatible work materials that are of low tensile strength, low-level heat generation, or high abrasiveness, such as:

1. Nonferrous metals: aluminum, babbit, brass, bronze, copper, silver, zinc, platinum
2. Nonmetals: Bakelite, berylia, glass, graphite, plastics
3. Composites: asbestos, carbons, Fiberglass, Nylon, PVC, Teflon

When machining ferrous materials with diamond cutting tools, the chips produced are tenacious, generating pressures that induce the chipping of cutting edges. In addition, a chemical reaction under high cutting temperatures can cause diamond to break down to its original graphite form.

Cubic Boron Nitride (CBN) is less expensive than diamond but much more expensive than carbide. However, CBN tools will last more than fifty times longer than carbide tools. Typical applications for CBN are:

1. Chilled cast iron
2. Carburized nitride or flame hardened parts
3. Hardened forged steel rolls (45–65 HRc)
4. Abrasion resistant parts (55–65 HRc)
5. High-speed steel tools of approximately 60 HRc

CBN cutting tools can tolerate a high degree of mechanical shock produced by interrupted cuts, such as when machining castings. Typically, CBN tools are used at speeds and feeds two to three times higher than those achievable with carbide tooling. The higher cutting speeds and feed rates, together with the longer tool life, result in lower machine and labor costs, which compensate for higher initial tool costs.

Carbide Inserts Classification

Cemented carbide inserts for special tool holders provide the widest applications in manufacturing. These inserts are indexable (Figure 6–6).

Figure 6–6 Indexable cemented carbide inserts provide the widest applications in manufacturing. (Courtesy Valenite, Inc.)

There is a wide range of carbide grades and insert shapes for almost any application. Most manufacturers comply with the American National Standards Institute (ANSI) and the International Standards Organization (ISO). ANSI allows manufacturers to create their own carbide grades and classifications. The ISO standard recommends a grading system that recognizes carbides according to application. The letters P, M, and K stand for different carbide grades. Each is assigned a corresponding color code as follows:

P (blue): Long chipping materials; steel, steel castings, and martensitic/ferritic stainless steels
M (yellow): austenitic stainless steel, superalloys, and titanium
K (red): short chipping materials: cast iron, hardened materials, and nonferrous materials

To improve productivity and reduce manufacturing costs, it is very important that the programmer and the operator have a thorough knowledge of carbide inserts and their applications. It is a time-consuming process to change the inserts at random in order to select proper grades for machining various materials. To help this situation, manufacturers have developed catalogs that greatly simplify the selection and use of cutting tools.

As illustrated in Figure 6–7, some manufacturers use a laser to permanently mark the catalog number on the insert; the grade identity, machining application (F = finishing, L = light, M = medium, R = roughing, and H = heavy), and the size of the tool nose radius are marked by sintering.

There are several factors that influence selection of the proper insert for an application. Programmers and operators should know these factors:

1. Operating conditions
2. Carbide grade
3. Insert size
4. Insert shape
5. Tool style
6. Tool nose radius

Figure 6–7 The insert is permanently marked for grade (P = Steel), application (F = finishing), tool nose radius (0.8 mm), and catalog number (4015). This information cannot wear off. (Courtesy Sandvik Coromant Company.)

The **operating conditions** are finishing, light, medium, roughing, and heavy machining; the carbide grades are P, M, or K. The **insert size** depends on operating conditions. For roughing and heavy machining, larger inserts are used.

The insert shape depends on tool style. Inserts are available in a number of shapes, such as square, triangle, rectangle, diamond, circle, or hexagon. They are made to have a number of indexable cutting edges. The standard insert shapes are square, triangular, and round. **Round inserts** provide the greatest strength and number of cutting edges; however, because the tool nose radius is often larger than the part radii, they are limited to a straight turning application. **Square inserts** are not as strong as round inserts, but are stronger than triangular inserts. **Triangular inserts** provide the most versatility. **Diamond-shaped inserts** are also very popular in finish turning applications since they have two cutting edges with acute angles that allow machining in narrow spaces. Square- and **hexagon-shaped inserts** are used most often in, but are not limited to, milling operations.

The choice of a tool style depends on a particular operation and operating conditions. A tool holder/insert combination should be chosen according to the specifics of the machine being used, available workholding device, and the machining operation being performed.

Tool nose radius depends on the surface finishing required and part geometry. For example, a $^1/_{32}$ tool nose radius cannot be used to make a $^1/_{64}$ radius on the part. Also, a bigger tool nose radius produces better surface finishing and a longer tool life.

Cutting Fluid

When metal is being machined, the friction between the tool and the part produces heat. The heat can quickly change the properties of the cutting tool material and dull the cutting edges. To reduce the heat and increase tool life, a **cutting fluid** is applied. A cutting fluid allows higher speeds and feeds to be used in machining. It can also be used to help remove chips from the cut. Thus, the cutting fluid plays an important role in how efficiently metal is removed in a machining operation. As illustrated in Figure 6–8, cutting fluid is most effective when applied generously at the exact point where the tool cutting edge penetrates the part.

Soluble oil and water are the most common cutting fluids used, while mineral- and sulphur-based oils can also be used. The synthetic-based cutting fluids offer better cooling, while oil-based solutions provide better surface finish. Cast iron, brass, and plastics are generally cut **dry,** meaning compressed air can be used, if needed, to cool the cutting tool. When cutting other materials, a cutting fluid is normally applied. Cutting fluid should always be used when cutting exotic metals. Water-soluble oil is suggested, but longer tool life can be achieved with sulfurized oil.

Figure 6–8 Cutting fluid being applied at the exact point where the tool cutting edge penetrates into the part. (Courtesy Valenite, Inc.)

The use of either cutting fluid or compressed air ensures adequate clearing of chips and prevents the recutting of chips. Recutting of chips reduces tool life, damages cutting edges, and affects the surface finish.

HSS tools will almost always perform better with cutting fluid; tools with carbide inserts may be run either dry or with cutting fluid. However, if cutting fluid is required, it must be applied generously. Operators should avoid interruptions in applying a cutting fluid, which can cause thermal shock that can result in damage to the tool cutting edge.

Operating Conditions and Tool Life

Operating conditions control the metal removal rate and tool life. **Metal removal rate** is the rate at which material is removed from the part. The operating conditions that determine the metal removal rate for an application are cutting speed, feed rate, and depth of cut.

Cutting speed is the speed at which a point of the edge of the cutting tool travels in relation to the part being machined. The cutting speed indicates that either the tool is moving past the part or the part is moving past the tool. Recommended cutting speeds for machining various materials are the result of research and practical experience, and can be found in engineering reference books or obtained from material manufacturers. Some cutting tool manufacturers also provide cutting data on labels on the insert boxes, making it easier to program and check cutting data on the shop floor (Figure 6–9).

The **depth of cut** may be defined as the depth of the chip removed by the cutting tool in one pass, measured in inches. On the lathe, the depth of cut is programmed as the radius value, known as *single depth of cut.*

Feed rate is the rate at which the tool moves into the part being machined. In milling applications the feed rate is referred to as feed per tooth, while in turning applications feed per spindle revolution is normally used. Feed rate can be stated in inches per minute or inches per revolution, as discussed earlier in the text.

Whenever the cutting speed, feed rate, or depth of cut is increased, the metal removal rate increases. If any of the operating conditions are decreased, the metal removal rate is reduced. Also, an equal change in any of the operating conditions will have an equal effect on the metal removal rate.

However, each of the three operating conditions has a different effect on tool life. A change in any of the operating conditions by an equal amount will not have the same effect on tool life. Cutting speed has the largest effect on tool life. If it is increased by 50 percent, tool life can be reduced almost 100 percent. Because of the great impact of cutting speed on tool life, the determination of cutting speed is the most critical factor when establishing the operating conditions. Cutting speed should be selected to maximize production rate or to minimize cost per piece. The operator cannot achieve both at the same time. Increasing cutting speed reduces machining time and consequently machining costs, but it also reduces tool life and increases

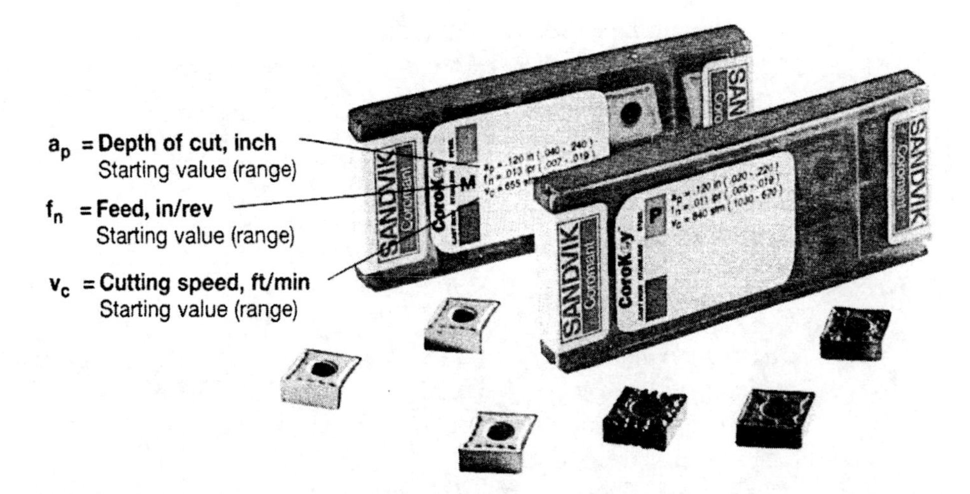

Figure 6–9 Insert-specific cutting data printed on insert box labels provides recommended starting cutting data and range values. (Courtesy Sandvik Coromant Company.)

cutting tool and tool change costs. The maximum production rate is achieved by selecting a cutting speed that best balances metal removal rate and tool life.

The feed rate has more impact on tool life than the depth of cut. In general, tool life decreases by the same percent as the feed rate increases. On the other hand, even a significant change in depth of cut reduces tool life only a small percentage.

Sometimes the cutting speed, feed rate, or depth of cut has to be adjusted according to a particular material or the way of holding the part. This should be done in such a way that the production rate stays the same. For instance, if the operator must reduce the cutting speed, he or she must increase the feed rate or depth of cut to keep the same production rate. In general, a larger depth of cut can be used for a slower speed, and surface finish will not be greatly affected. If the speed is decreased and the feed rate increased, the surface finish will be affected. When adjusting the cutting conditions, do not change the speed and feed rate at the same time, since it can lead to confusion.

As mentioned earlier, some cutting tool manufacturers provide cutting data on labels attached to the cover of the dispenser (Figure 6–10). Cutting data recommendations—a starting value and a total range value—provide a quick and easy start on the shop floor. The programmer can enter recommended values or conservative values from the range value for an application. Then, if everything looks good when machining, he or she can increase these values using the higher values from the range value. This will increase productivity, since the programmer and the operator do not have to guess or look up information.

When machining, always select the largest insert radius that the part design will allow. An insert with a larger radius is stronger because it has more support for the insert corner. It will also better dissipate the heat while cutting. Thus, an insert with a larger tool radius can take heavier cuts as well as higher cutting speeds, which will improve the production rate. An insert with a larger tool radius also produces better surface finishing than one with a small tool radius.

Adaptive Control

Adaptive control is a control technology related to CNC machining, which is still in the development stage. Adaptive control allows the machine control to adjust cutting conditions during machining. It is the automatic response of the control based on the parameter setting; the operator does not have control over this function.

In adaptive control, the machine control monitors and compares the cutting torque with the torque limits (high and low) set by parameter setting. If the cutting torque is lower than the low torque limit, the control will increase the feed rate or cutting speed. If the cutting torque exceeds the high torque limit, the control will decrease the feed rate or cutting speed, or even stop the cycle.

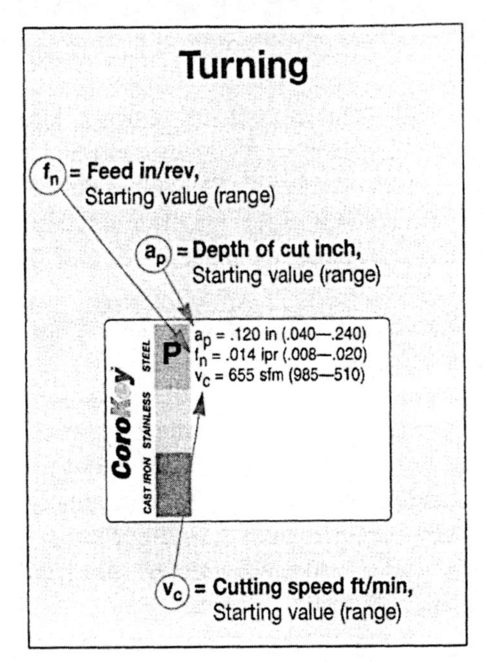

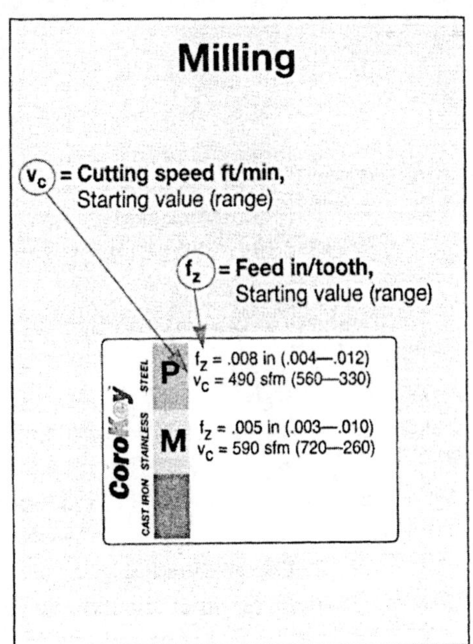

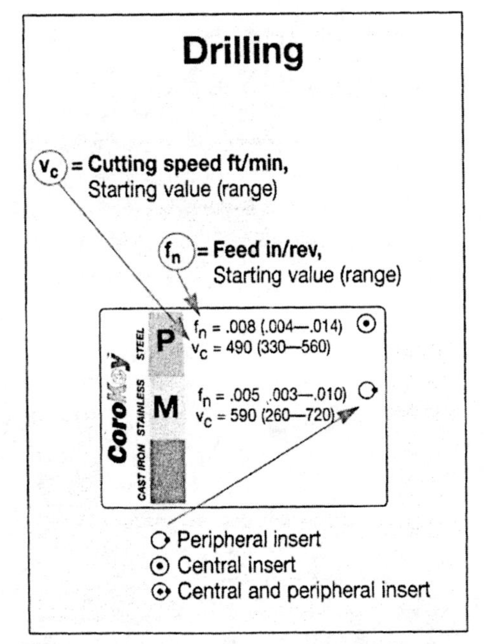

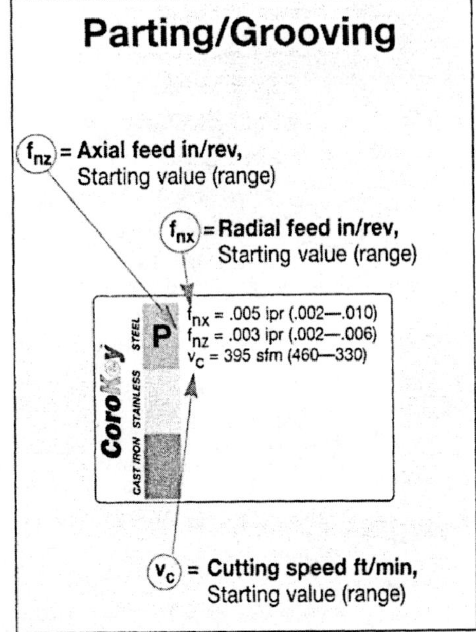

Figure 6–10 Cutting data labels on insert boxes make it easy to program or check data for an application. (Courtesy Sandvik Coromant Company.)

Adaptive control provides real-time adjustments of cutting conditions. It reacts on higher values as well as on lower values programmed for cutting speeds and feed rates, improving production rates from 20 to 80 percent. The improvement in production rates is more significant when machining complex parts, since programmers tend to enter conservative values in such cases.

Summary

CNC machine tools are made to improve productivity and reduce manufacturing costs. Selecting the right cutting tool is essential to obtaining the best results. There are two major types of cutting tools used on CNC machines, single-point tools and multipoint tools. Single-point tools are used primarily in turning and boring operations on turning centers, but they are also used for boring on machining centers. In modern metal-cutting operations, all single-point cutting tools use indexable carbide inserts. They are available in a variety of shapes.

Multipoint tools, such as milling cutters, end mills, shell end mills, and reamers, are used for milling, drilling, reaming, counterboring, and countersinking operations. Smaller multipoint cutting tools have two or four cutting edges, while larger sizes have more than four.

In modern manufacturing, the following materials for cutting tools are used most often:

1. High-speed steel (HSS)
2. Cemented carbides
3. Ceramics
4. Diamond
5. Cubic Boron Nitride (CBN)

High-speed steel cutting tools are used when cutting with lower cutting speeds and lower tool contact temperatures. HSS tools can be easily resharpened, which is their main advantage.

Cemented carbide tools are used when cutting with higher cutting speeds and higher tool contact temperatures. Because of higher cutting speeds and longer tool life, carbide tools are widely accepted as a cutting tool material for milling and turning applications.

Ceramics are cemented oxide tool materials made by sintering in a similar fashion as cemented carbides. They are harder than carbides, and allow cutting at higher temperatures and cutting speeds.

Synthetic diamond, known as polychrystaline-diamond, or PCD, is more suitable for cutting tools than natural diamond. Polychrystaline-diamond tooling is often used in precision machining and finish operations of nonferrous (nonmetalic) materials.

Cubic Boron Nitride (CBN) is less expensive than diamond, but much more expensive than carbide. However, CBN tools will last more than fifty times longer than carbide tools. These tools are used at speeds and feeds two to three times higher than those achievable with carbide tooling.

There is a wide range of carbide grades and insert shapes for almost any application. The ISO standard recommends a grading system that recognizes carbides according to application. The letters P, M, and K are assigned to a grade and a corresponding color code:

P (blue): Long chipping materials: steel, steel castings, and martensitic/ferritic stainless steels
M (yellow): austenitic stainless steel, superalloys, and titanium
K (red): Short chipping materials: cast iron, hardened materials, and nonferrous materials

Inserts are available in a number of shapes, such as square, triangular, rectangle, diamond, circle, or hexagonal. They are made to have a number of indexable cutting edges. To improve productivity and reduce manufacturing costs, it is very important that programmers and operators have a thorough knowledge of carbide inserts and their applications.

When metal is being machined, the friction between the tool and the part produces heat. The heat can quickly change the properties of cutting tool material and dull the cutting edges. To reduce the heat and increase the tool life, a cutting fluid is applied. This fluid allows higher speeds and feeds to be used in machining.

Soluble oil and water are the most common cutting fluids, but mineral- and sulphur-based oils can also be used. Synthetic-based cutting fluids offer better cooling, while oil-based solutions provide better surface finish.

Cast iron, brass, and plastics are generally cut dry, meaning compressed air is used to cool the cutting tool. When cutting other materials, a cutting fluid is normally applied. Cutting fluid should always be used when cutting exotic metals.

Operating conditions control metal removal rate and tool life. Metal removal rate is the rate at which material is removed from the part. The operating conditions that determine the metal removal rate for an application are cutting speed, feed rate, and depth of cut.

An equal change in any of the operating conditions will have an equal effect on the metal removal rate. However, each of the three operating conditions has a different effect on tool life.

Cutting speed has the largest effect on tool life. Because of the great impact of cutting speed on tool life, the determination of cutting speed is the most critical factor when establishing the operating conditions.

The feed rate has more impact on tool life than depth of cut does. In general, tool life decreases by the same percentage as the feed rate increases. On the other hand, even a significant change in depth of cut reduces tool life only a small percentage.

When machining, an operator should always select the largest insert radius that the part design will allow. An insert with a larger radius is stronger because it offers more support for the insert corner. It will also better dissipate the heat while cutting. Thus, an insert with a larger tool radius can take heavier cuts as well as higher cutting speeds, which will improve the production rate. An insert with a larger tool radius also produces better surface finishing than one with a small tool radius.

Adaptive control allows the machine control to adjust cutting conditions during machining. It is the automatic response of the control based on the parameter setting. Adaptive control provides real-time adjustments of cutting conditions. It reacts on higher values as well as on lower values programmed for cutting speeds and feed rates, improving production rates from 20 to 80 percent.

Key Terms

adaptive control
cemented carbides
ceramics
coated carbides
Cubic Boron Nitride (CBN)
cutting fluid
cutting speed
depth of cut
diamond-shaped inserts
feed rate
hexagon-shaped inserts
high-speed steel (HSS)
indexable inserts
metal removal rate
multipoint tools
operating conditions
polychrystaline-diamond tooling
single-point tools
square inserts
round inserts
triangular inserts

Self-Test

Answers are in Appendix E.

1. The _______________ are used primarily in turning and boring operations on turning centers.
2. _______________ are used for milling, drilling, reaming, counterboring, and countersinking operations.
3. _______________ cutting tools are used when cutting with lower cutting speeds and lower tool contact temperatures.
4. _______________ tools are used when cutting with higher cutting speeds and higher tool contact temperatures.
5. _______________ are used for machining hard materials with high speeds using a small depth of cut.
6. _______________ is often used in precision machining and finish operations of nonmetallic materials.
7. Typically, _______________ tools are used at speeds and feeds two to three times higher than those achievable with carbide tooling.

8. The _______________ provide the greatest strength and number of cutting edges.
9. The _______________ are not as strong as round inserts, but are stronger than triangular inserts.
10. The _______________ provide the most versatility.
11. The _______________ are very popular in finish turning applications. They have two cutting edges with acute angles that allow machining in narrow spaces.
12. The _______________ are mostly used in milling applications.
13. _______________ is applied to reduce heat and increase tool life.
14. The _______________ controls the metal removal rate and tool life.
15. _______________ is the rate at which material is removed from the part.
16. The _______________ is the speed at which a point of the edge of the cutting tool travels in relation to the part being machined.
17. _______________ is the rate at which the tool moves into the part being machined.
18. _______________ is the automatic response of the control system that allows adjustment in cutting conditions during machining.

Relating the Concepts

No answers are suggested.

1. Name two major types of cutting tools. Indicate their use.
2. List the cutting tool materials.
3. What are the advantages and disadvantages of cemented carbide tools?
4. Explain the ISO grading system for carbide tools.
5. Which factors influence selection of a proper insert for an application?
6. Why is cutting fluid applied when machining?
7. Explain how cutting speed, feed rate, and depth of cut affect the metal removal rate.
8. Explain how cutting speed, feed rate, and depth of cut affect tool life.
9. Why is selecting the largest insert radius when machining recommended?
10. What is adaptive control? How does it work?

7 Programming CNC Lathes

Key Concepts

Lathe Canned Cycles

Diameter Cutting by the G71 Cycle

Face Cutting by the G72 Cycle

Pattern Repeating by the G73 Cycle

Finish Cutting by the G70 Cycle

Threading

- Programming Threading in a Single Block by the G76 Cycle
- G32, the Most Versatile Canned Cycle
- The G92 Cycle, the Accelerated G32 Cycle
- Cutting Deep or Wide Threads
- Cutting Left-Hand Threads
- Cutting Multistart Threads

Grooving

- Diameter Grooving by the G75 Cycle
- Face Grooving by the G74 Cycle

Lathe Canned Cycles

Basic CNC lathes are two-axis machines; the Z axis is line parallel to the spindle centerline, and the X axis is perpendicular to it. More advanced CNC lathes are known as turning centers, equipped with four or more axes that allow turning, drilling, and milling in one setup. Most CNC turning centers (Figure 7–1) are designed for machining shaft-type parts that are held in a chuck and supported by a tailstock center. The tools are held in a turret with eight to twelve stations. To make the tool change, the programmer can assign the turret to rotate forward, reverse, or in the closest direction.

To simplify the programming task and save time, CNC machine builders have designed built-in programs that are stored in the CNC machine electronic circuits for permanent use; these are known as **canned cycles.** There are canned cycles for face and diameter roughing, finishing, profile copying, grooving on face and diameter, threading, and so on. Each canned cycle has a set of rules that the programmer can follow easily by entering such constants as depth of cut, finishing allowance, drilling depth, and thread pitch. For example, one lathe canned cycle allows cutting any depth of thread with only a single line of information.

Lathe canned cycles are divided into two groups: fixed cycles and multiple repetitive cycles. **Fixed cycles** permit the programmer to program three or four successive tool movements in a single block. They also permit the repetition of operations with a small change in the next block. Following are fixed cycles:

G32 Threading cycle
G90 Diameter cutting cycle
G92 Threading cycle
G94 Face cutting cycle

Note that the G90 and G94 fixed cycles are not used much since the more advanced canned cycles have replaced them.

Figure 7–1 CNC turning center, which is designed for maximum efficiency in turning and boring operations, allowing a quick tool change in seconds. (Courtesy Hitachi Seiki U.S.A., Inc.)

Multiple repetitive cycles, also known as automatic repeat cycles, are more sophisticated than fixed cycles. Using them, the programmer can repeat any number of passes in a relatively short program; the tool cuts the material in the repeating sequences until the specified profile is achieved. There is no limit to the number of repeat passes: The control calculates them according to specified data. Each of these cycles has its particular purposes, but they are all similar for programming. On the CNC lathe, the following repetitive canned cycles are used most often:

G71 Diameter roughing cycle
G72 Face roughing cycle
G73 Pattern repeat cycle
G70 Finishing cycle
G74 Face peck-grooving cycle
G75 Diameter peck-grooving cycle
G76 Threading cycle

Diameter Cutting by the G71 Cycle

When roughing large amounts of material in turning operations, the tool changes its position in each of the many passes. Programmers use canned cycles or part programming software to make programming these changes simpler. For diameter cutting along the Z axis, either turning or boring, the G71 canned cycle is most often used. The programmer describes the finish part profile between the P and Q blocks. The control then determines the number of passes, depending on the amount of material to be removed from the outside diameter when turning, or from the hole diameter when boring. Taking into consideration the depth of cut, the control calculates the number of passes based on two things:

1. The tool start point, which is slightly above the outside stock diameter when turning, or slightly below the inside stock diameter when boring
2. The part finish profile

After roughing is completed, the tool leaves a specified amount of material for a finishing cycle that can be removed by the same or a different tool. Figure 7–2 shows the basic structure of the G71 cycle, which is illustrated below.

...; (Commands for coordinate system preset, tool change, etc.)
G00 X_ Z_; (Positioning to point A.)
G71 P__ Q__ U__ W__ D__ F__ S__ T__; (The G71 cycle is specified.)
N... G00 X__; (The tool moves to point B, the cycle start point specified by the address P. The block number must be programmed.)
...; (Cutting the part profile)
N...; (The tool moves to point C; the cycle end point specified by the address Q. The block number must be programmed.)
...; (Return home or start of finishing cycle.)

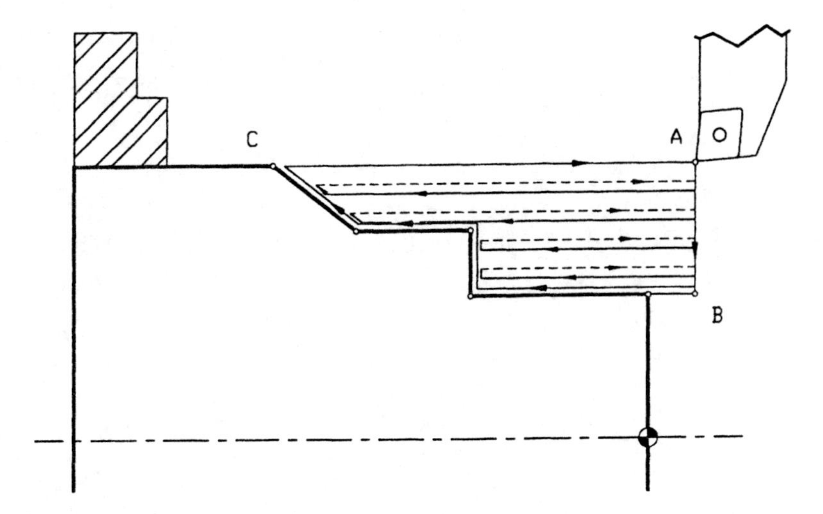

Figure 7–2 Basic structure of the G71 cycle.

Address Description:

- P The block number indicating the cycle start point. This is the first block after the G71 cycle is specified, instructing the tool to move to the first point on the finish part profile.
- Q The block number indicating the cycle end point, the last point on the finish part profile.
- U Finishing allowance on the X axis (diameter value) with a plus sign when turning or a minus sign when boring.
- W Finishing allowance on the Z axis, specified with a plus sign for ordinary cutting or a minus sign for back cutting.
- D The depth of cut (radius value), specified with a positive sign only. On some controls the decimal point format cannot be used, so the programmer must assume four decimals in leading zeroes supression format. For example, for a 0.1 inch depth of cut, he or she would enter 1000.
- F Feed rate.
- S Spindle speed that will come in effect in the finishing cycle if programmed between the P and Q block numbers.
- T Tool offset number that will come in effect in the finishing cycle if programmed between the P and Q block numbers.

Before entering the canned cycle, the programmer must bring the tool to the *tool start position* (point A in Figure 7–2), which is the highest tool X position when outside diameter cutting or the lowest tool X position when inside diameter cutting. (After the cycle is finished, the tool returns to this point.)

Now the programmer must describe the part profile. First, he or she uses G00 or G01 to program the first block after the G71 block with the motion on the X axis only, moving the tool to the *cycle start position* (point B). This is the lowest tool X position when turning or the highest tool X position when boring. This point has the same Z coordinate as point A. The block number must be specified by the address P. Then the programmer moves the tool along the part finish profile until it reaches point C, the last point in the canned cycle. Usually, point C has the same X coordinates as point A. At this point, the programmer must enter a block number by the address Q. When the control reads this code, it knows that the cycle is finished; the tool retracts to point A. (The programmer does not have to reenter the coordinates of point A or cancel the cycle by the G00 code, but it will not make any difference if programmed.)

The tool does not cut the part following the part profile directly, as specified by the programmer. This would damage the tool and possibly the machine. Instead, the tool cuts the steps along the Z axis by the specified feed rate. The first step is from point A to near point C. The depth of each step depends on the programmed depth of cut specified by the address D. The control calculates the Z coordinate in order to know where to stop the tool to avoid hurting the finish part profile. After each pass along the Z axis, the tool feeds out from the part on the X axis. Then the tool returns in rapid to the start position (point A). The control repeats this process for each pass while roughing through the following pattern:

Rapid down on X to take a cut.
Cutting along the Z axis by the programmed feed rate.
Feeding out on the X axis by the same feed rate.
Rapid back on Z to the start position.

After each pass on the Z axis, the tool does not return to the X coordinate of point A. It retracts from the part diameter being cut for the amount specified by parameter, speeding up the process. The retract amount is usually 0.010 inch, but it may be more or less.

After roughing out all of the steps, the tool leaves some amount of material (also set by parameter) for the last pass. Then, the tool moves to point B and cuts the finish profile up to point C. If the programmer does not want the finishing cycle, the part is complete; otherwise, some amount of material (specified by the addresses U and W) is left for the finishing cycle. When the programmer does not want the finishing cycle, he or she omits the addresses U and W or programs zero values. The control assumes that there is no finishing allowance and calculates only the number of roughing passes.

The feed rate specified in the G71 block will be in effect for all the roughing passes. To achieve a variation in surface finish, the programmer must use the finishing cycle. Then, he or she should program different feed rates and spindle speeds inside the roughing cycle; they will come into effect in the finishing cycle.

The G71 cycle has four different patterns (Figure 7–3). The X coordinates must successively increase or decrease in each of these patterns.

The ordinary way of cutting the outside part diameter is illustrated at position (a). Back cutting on the outside part diameter is shown at position (b). Positions (c) and (d) show the same machining operations performed on the inside part diameter.

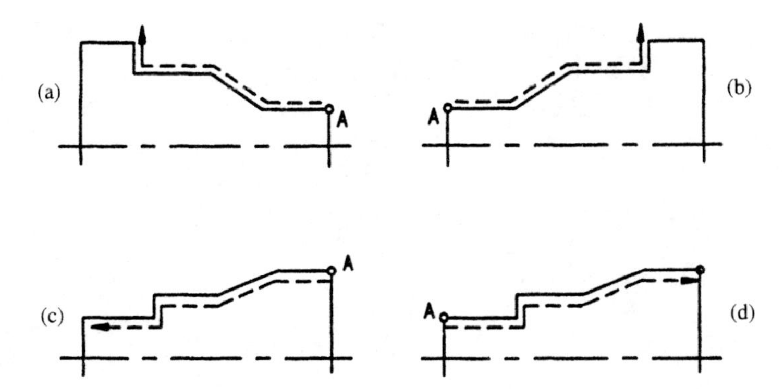

Figure 7–3 Cutting patterns for the G71 cycle.

If a profile with pocket or undercut is to be cut, the programmer should select a modified G71 cycle (Figure 7–4).

When using a modified G71 cycle, the programmer uses G00 or G01 to program the first line after the G71 block with motion on both the X and Z axes. For example:

```
G00 X__ Z__;
G01 X__ Z__;
```

Following is a programming example using the G71 cycle (Figure 7–5). The material is 4140 steel and the stock size is $4\frac{1}{8}$ inches.

O100; (Program number.)
G50 X15.0 Z3.0 S1500 M42; (Coordinate system preset.)
G00 T0101; (Tool and tool offset call.)
G96 S390 M03; (Spindle start; constant surface speed.)
G00 X4.125 Z0.1 M08; (Rapid to starting position.)
G71 P10 Q20 U0.012 W0.003 D1300 F0.013; (Cycle starts at line 10 and finishes at line 20 as specified by the P and Q addresses. The finishing allowance is 0.012 inch on the X axis and 0.003 inch on the Z axis, as entered by the U and W addresses. The depth of cut of 0.13 inch is specified by the address D using the leading zeroes supression format. The feed rate of 0.013 inch per revolution is specified by the address F.)

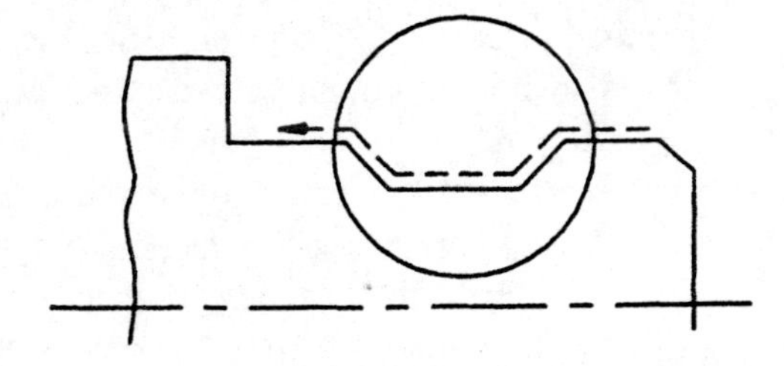

Figure 7–4 A modified G71 cycle is used to cut the part profile with undercut.

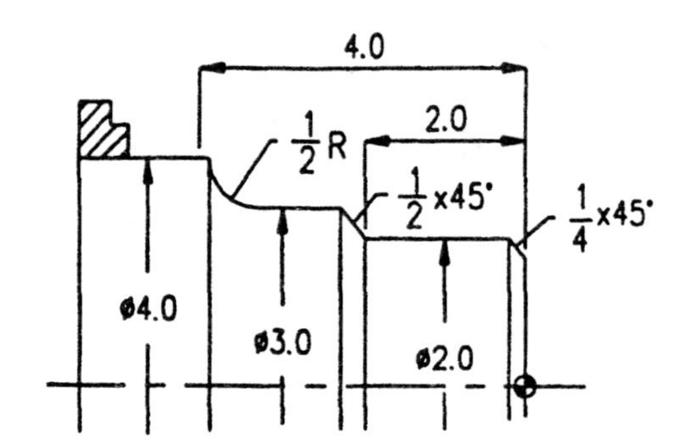

Figure 7–5 Programming the G71 cycle.

N10 G00 X1.5; (Cycle start point specified by the address P; the line number must be programmed. This is the move in rapid on the X axis only. The X coordinate specified is the lowest tool position on the part finish profile.)
G01 G42 Z0 F0.015; (Feeding to a Z0. Compensation right by the G42 code.)
X2.0 Z-0.25 F0.009; (Chamfer cutting.)
Z-2.0 F0.012; (Straight cutting.)
X3.0 W-0.5 F0.009; (Chamfer cutting.)
Z-3.5 F0.012; (Straight cutting up to the radius start point.)
G02 X4.0 Z-4.0 R0.5 F0.008; (Radius cutting.)
N20 G01 Z-4.2 F.012; (Straight cutting. The last point of the part profile is specified by the address Q. This is the cycle end point. The line number must be programmed.)
G00 G40 X15.0 Z3.0 M09; (Return to the home position. Compensation cancel.)
T0100; (Tool offset cancel.)
M01; (Program temporary stop.)

When reading line 20, the control knows that this is the last point of the part profile. After executing the line, the control returns the tool to the cycle start point, X4.125 Z0.1. At this point, the programmer should not enter the M code or line number since on some controls it may repeat the cycle. To avoid this, the programmer should enter the G code by programming the finishing cycle, returning the tool to home position, or positioning the tool to another location.

In this program, the different feed rates are entered according to the specifics of the part profile. If the finishing cycle is to be used, the variation in the feed rate will come into effect. If the finishing cycle is not needed, the part will be completed by the roughing feed rate specified in the G71 block. The G42 code is programmed after the G71 block, enabling the use of the tool radius compensation for the finishing cycle. Before returning the tool to home position, the compensation must be canceled, no matter if the finishing cycle will be used or not.

On the latest controls, the G71 cycle is programmed in two blocks. The retract amount by the R address and depth of cut by the U address are specified in the first block. The rest of the data is entered in the second block, as illustrated in the following series of instructions:

G00 X4.125 Z0.1; (Rapid to the start position.)
G71 U0.13 R0.05; (First part of the G71 command.)
G71 P10 Q20 U0.012 W0.003 F0.012; (Second part of the G71 command.)
N10 G00 X1.5; (First point in the G71 cycle is specified.)
...; (Continuing.)

The U address specified in the first part of the G71 command expresses the depth of cut, while in the second part of this command it represents the finishing allowance on the X axis. This way of programming the G71 cycle enables the programmer to control the tool retract amount by the address R rather then depend on parameter setting. This is an important consideration when taking the first cut with a slightly oversized boring bar; if needed, the operator can quickly decrease the retract amount, preventing the tool from touching the part with its back side.

Face Cutting by the G72 Cycle

Programmers use the G72 cycle for facing and roughing parts that are larger in diameter than in length. In this cycle the tool cuts the part along the X axis. The finish part profile is specified between the P and Q blocks; then the control calculates the number of passes required depending on the amount of material to be removed. After the roughing is completed, the tool leaves a specified amount of material for the finishing cycle, which can be removed by the same or a different tool. Following is the basic structure of the G72 cycle (Figure 7–6):

...; (Commands for coordinate system preset, tool change, etc.)
G00 X_ Z_; (Positioning to point A.)
G72 P__ Q__ U__ W__ D__ F__ S__ T__; (The G72 cycle is specified.)
N... G00 Z__; (The tool moves to point B, which is the cycle start point specified by the address P. The block number must be programmed.)
...; (Cutting the profile.)

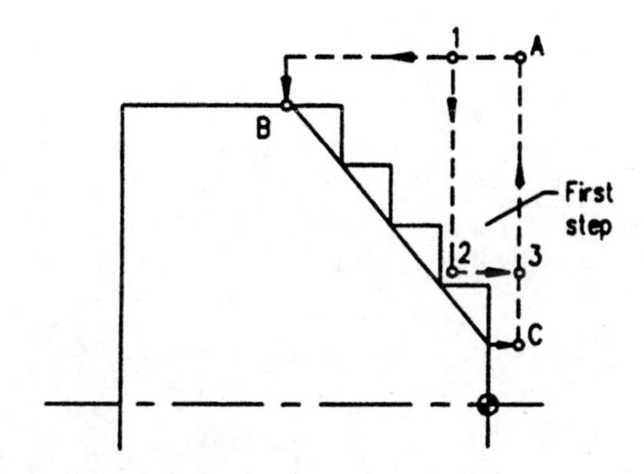

Figure 7–6 The structure of the G72 cycle.

N...; (The tool moves to point C, which is the cycle end point specified by the address Q. The block number must be programmed.)
...; (Positioning to another point, return home, or start the finish cycle.)

Address Description:

P The block number indicating the cycle start point. This is the first block after the G72 cycle is specified, instructing the tool to move to the first point on the finish part profile.

Q The block number indicating the cycle end point; the last point on the finish part profile.

U Finishing allowance on the X axis (diameter value), with a plus sign when turning or a minus sign when boring.

W Finishing allowance on the Z axis, specified with a plus sign for ordinary cutting or a minus sign for back cutting.

D The depth of cut (radius value) specified with a positive sign only. On some controls decimal point format cannot be used, so the programmer must assume four decimals in leading zeroes supression format. For example, for a 0.12 inch depth of cut, he or she would enter 1200.

F Feed rate.

S Spindle speed that will come into effect in the finishing cycle if programmed between the P and Q block numbers.

T Tool offset number that will come into effect in the finishing cycle if programmed between the P and Q block numbers.

Before entering the canned cycle, the programmer must bring the tool to the *tool start position* (point A on Figure 7–6), the highest tool X position when turning. After the cycle is finished, the tool returns to this point.

The programmer describes the part profile by entering the first block after the G71 block with the motion on the Z axis only, using the G00 or G01, and moving the tool to the *cycle start position* (point B), the longest tool position on the Z axis. The block number must be specified by the address P. Then the programmer moves the tool along the part finish profile until it reaches point C, the last point in the canned cycle. Usually, point C has the same Z coordinates as point A. At this point, the programmer must enter a block number by using address Q. When the control reads this code, it knows that the cycle is finished; the tool retracts to point A.

The tool cuts the steps along the X axis by the specified feed rate. The first step is from point A to near point C. The depth of each step depends on the programmed depth of cut specified by the address D. The control calculates the X coordinate in order to know where to stop the tool to avoid hurting the finish part profile. After each pass along the X axis, the tool feeds out from the part on the Z axis. Then, the tool returns in rapid to the start position (point A). The control repeats this process for each pass while roughing through the following pattern:

Rapid on the Z axis to take a cut.
Cutting along the X axis by the programmed feed rate.

Feeding out on the Z axis by the same feed rate.
Rapid back on X to the start position.

After roughing out all the steps, some amount of material is still left for the last pass. Then the tool moves to point B and cuts the finish profile up to point C. If the programmer does not want the finishing cycle, the part is complete; otherwise, some amount of material (specified by the addresses U and W) is left for the finishing cycle. When the programmer does not want the finishing cycle, he or she omits the addresses U and W or programs zero values.

The feed rate specified in the G71 block will be in effect for all the roughing passes. To achieve a variation in surface finish, the programmer must use the finishing cycle. Then, he or she should program different feed rates and spindle speeds inside the roughing cycle; they will come into effect in the finishing cycle.

The following is a practical example of programming the G72 cycle (Figure 7–7). The material is 4140 steel and the stock size is 5⅛ inches.

```
G50 X15.0 Z3.0 S1300 M42; (Coordinate system preset. Speed limit. High
    range.)
G00 T0101; (Tool and tool offset call.)
G96 S395 M03; (Spindle start.)
G00 X5.2 Z0.1 M08; (Rapid to position. Coolant on.)
G72 P30 Q50 U0.015 W0.005 D1200 F0.011; (The G72 cycle is specified.)
N30 G00 Z-1.5; (Cycle start point specified by the P address; the move on the Z
    axis only.)
G01 G41 X5.0 F0.02; (Approaching by a faster feed. Compensation left by the
    G41 code.)
X4.0 Z-1.0 F0.009; (Chamfer cutting.)
Z-0.75 F0.013; (Straight cutting.)
G02 X3.5 Z-0.5 R0.25 F0.008; (Radius cutting.)
G01 X3.0 F0.012; (Shoulder cutting.)
Z-0.125; (Straight cutting.)
X2.75 Z0; (Chamfer cutting.)
N50 Z0.1 F0.015; (Cycle end specified by the address Q; the tool feeds out on
    the Z axis.)
```

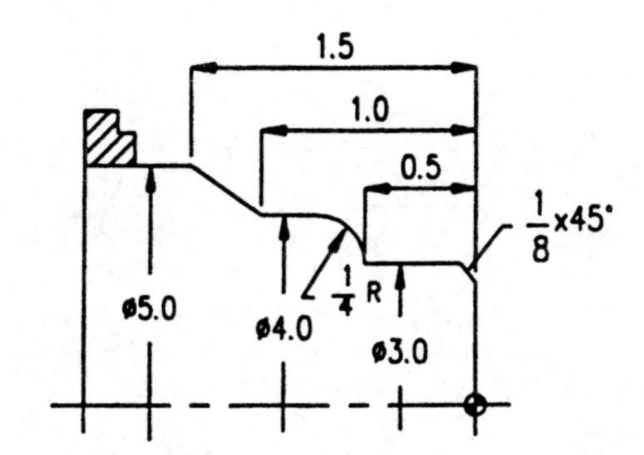

Figure 7–7 Programming the G72 cycle.

G00 G40 X15.0 Z3.0 T0100 M09; (Return home. Compensation cancel. Offset cancel.)
M01; (Optional program stop.)

The G41 code, which is programmed after the canned cycle is specified, calls for tool radius compensation to be used for the finishing cycle. Before returning the tool to the home position, the compensation must be canceled, whether the finishing cycle will be used or not.

Pattern Repeating by the G73 Cycle

The G71 and G72 cycles are not efficient to use when roughing parts with a profile already preformed by casting or forging. Time will be lost when the tool makes a number of unnecessary passes cutting air, making the machining time the same as if the part was not preformed. To avoid this time loss, programmers use the G73 cycle with copying capabilities.

The G73 cycle is derived from the G71 cycle and works on the same principle. The only difference is that the tool is not roughing along the Z axis; it follows the preformed profile even when roughing, saving machining time. When programming the G73 cycle, the programmer specifies the pattern of the finish part profile between the P and Q block numbers. The control repeats the pattern a number of times until the finish part profile is cut. The number of passes depends on the finish part size and the depth of cut. Following is the instruction format:

G73 P__ Q__ I__ K__ U__ W__ D__ F__ S__ ;

Address Description:

P The block number indicating the cycle start point. This is the first block after the G73 cycle is specified, which instructs the tool to move to the first point on the finish part profile.

Q The block number indicating the cycle end point; the last point on the finish part profile.

I The amount of material to be removed on the X axis (radius value).

K The amount of material to be removed on the Z axis.

U Finishing allowance on the X axis (diameter value), with a plus sign when turning or a minus sign when boring.

W Finishing allowance on the Z axis, specified with a plus sign for ordinary cutting or a minus sign for back cutting.

D Depth of cut (radius value) on both the X and Z axis. Decimal point format cannot be used. On some controls, the depth of cut is expressed through the number of passes. The control then calculates the cutting depth according to the tool start point.

F Feed rate.

- S Spindle speed that will come into effect in the finishing cycle if programmed between the P and Q block numbers.
- T Tool offset number that will come into effect in the finishing cycle if programmed between the P and Q block numbers.

Following is a programming example for the part illustrated in Figure 7–8. The material is cast iron. The roughing allowance is 0.3 inch for the X axis and 0.05 inch for the Z axis. When programming the G73 canned cycle, the programmer must pay special attention to the sign for allowance on each axis, because there are four variations of this cycle (which are the same as the types of the G71 cycle). In this example, the sign for allowance on both axes is positive.

```
G50 X15.0 Z3.0 M41 S900; (Coordinate system preset. Speed limit. Low range.)
G00 T0101; (Tool and tool offset call.)
G96 S375 M03; (Spindle start; constant surface speed.)
G00 X7.1 Z0.15; (Positioning to the cycle start point.)
G73 P101 Q102 I0.3 K0.05 U0.015 W0.003 D1100 F0.012; (The G73 cycle is
    specified.)
N101 G00 X4.0; (The lowest X position on the part profile.)
G01 G42 Z0 F0.015; (Feeding to a Z0. Compensation right by G42 code.)
Z-1.0 F.012; (Straight cutting.)
X5.0 Z-1.5 F0.009; (Chamfer cutting.)
Z-2.5 F0.012; (Straight cutting.)
G02 X6.0 Z-3.0 R0.5 F0.009; (Radius cutting.)
G01 Z-4.0 F0.012; (Straight cutting.)
N102 X7.1; (Shoulder cutting. Cycle end.)
G00 G40 X15.0 Z3.0 T0100 M09; (Return home. Compensation cancel. Offset
    cancel.)
M01; (Optional program stop.)
```

Note that different feed rates are programmed inside the canned cycle; they will come into effect when the finishing cycle is executed, resulting in variation in

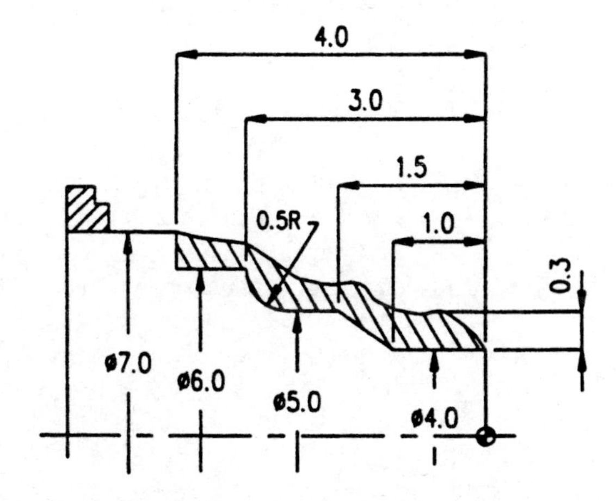

Figure 7–8 Programming the G73 cycle.

surface finishing. Also note that the G42 code is programmed after the G73 cycle is specified, enabling the programmer to use tool radius compensation for the finishing cycle. Before returning the tool to home position, compensation must be canceled, whether the finishing cycle will be used or not.

Finish Cutting by the G70 Cycle

Programmers program the G70 cycle after the G71, G72, or G73 multiple repetitive roughing cycles to cut the part profile to a finish size; the tool cuts the part in one pass, following the part profile specified in the roughing cycle. The depth of cut for the finishing pass depends on the finishing allowance programmed by the U and W addresses in the roughing cycle. Following is the instruction format when programming the finishing cycle:

G70 P__ Q__;

Address Description:

P The block number indicating the cycle start point. This is the first block after the roughing cycle is specified, which instructs the tool to move to the first point on the finish part profile.

Q The block number indicating the cycle end point; the last point on the finish part profile specified in the roughing cycle.

The block numbers designated by P and Q are the same block numbers programmed in the roughing cycle. This means that the programmer cannot affect the part profile by the use of the G70 cycle. If he or she wants any alteration in the tool path, feed rate, or cutting speed, he or she must use the roughing cycle. The P and Q block numbers must be in the memory of the CNC machine. If the control cannot find them, it raises an alarm.

Any variations in the feed rate and cutting speed programmed in the roughing cycle come into effect when the finishing cycle is executed. The programmer can call the finishing cycle immediately after the roughing cycle is finished using the same tool; then, the end point in roughing becomes the start point in finishing. Also, he or she can call the finishing cycle at any point later in the program, using the same tool for roughing or a different one. Usually, finish with a different tool is preferred so the size is held better.

To enable the control to use the automatic tool nose radius compensation in the G70 finish cycle, the programmer must enter the compensation call code inside the roughing cycle (between the P and Q block numbers). Following is an example of programming the finishing cycle for the part shown in Figure 7–9. The bar stock is 3 inches and the material is mild steel.

N1 G50 X15.0 Z3.0 S1200 M42; (Coordinate system preset. Speed limit. High range.)

G00 T0101; (Tool and tool offset call.)

Figure 7–9 Programming the G70 cycle.

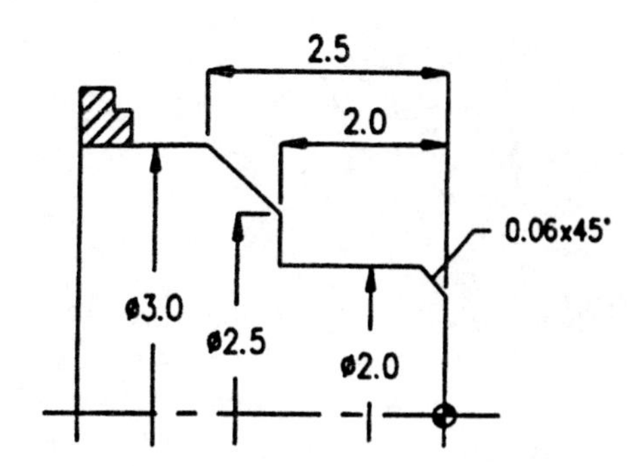

G96 S475 M03; (Constant surface speed. Spindle clockwise.)
G00 X3.0 Z0.1 M08; (Position before the roughing cycle. Coolant on.)
G71 P10 Q20 U0.015 W0.005 D1100 F0.012; (Roughing cycle is specified.)
N10 G00 X1.88; (Roughing cycle start block.)
G01 G42 Z0 F0.02; (Approaching by a faster feed. Compensation right.)
X2.0 Z-0.06 F0.009; (Chamfer cutting.)
Z-2.0 F0.012; (Straight cutting.)
X2.5 F0.01; (Shoulder cutting.)
X3.0 Z-2.5 F0.009; (Chamfer cutting.)
N20 X3.1 W-0.05 F130; (Roughing cycle end point. The tool is leaving the part at an angle to break the sharp edge.)
G40; (Compensation cancel for the roughing cycle.)
G70 P10 Q20; (Finishing cycle is specified.)
G00 G40 X15.0 Z3.0 T0100 M09; (Return home. Compensation cancel for the finishing cycle. Offset cancel. Coolant stop.)
M1; (Optional program stop.)

The roughing cycle is specified between lines N10 and N20. After executing line 20, the control returns the tool to the cycle start point X3.0 Z0.1, programmed in the block preceding the G71 block. From this point, the finishing cycle is executed.

The tool radius compensation must be canceled twice, after the roughing and after the finishing cycle. If it is not canceled after the roughing cycle, the compensation for the finishing cycle will build up; if it is not canceled after the finishing cycle, the next tool in the program will pick it up even if this tool does not need it. The programmer can enter the compensation cancel code in the Q block; then, the compensation for either cycle is canceled automatically. However, he or she should make sure that the tool has enough room to retract after canceling the compensation; otherwise, overcutting may occur.

In the example above, the finishing cycle starts immediately after the roughing cycle ends. When the programmer wants to use a different tool for the finishing, he or she returns the roughing tool home for the tool change. Then he or she brings the tool to the cycle start position (X3.0 Z0.1) and calls up the G70 cycle by entering G70 P10 Q20. In the next block, the programmer cancels the tool radius compensation and returns the tool home.

Threading

In thread cutting operations, the machine is in a special state that limits operator input during program execution. The operator cannot adequately stop the machine once it starts threading because the Feed Hold function usually has no effect. (Single-block execution *is* possible; the cutting tool stops after it returns to the cycle start point.) If the operator thinks that the tool might collide with the part or chuck, he or she must press the EMERGENCY STOP button. This means that the programmer and the operator must exercise special caution when programming and setting up the machine for threading operations.

The threading feed rates depend on the thread pitch or lead. They are much higher than turning or boring feed rates, causing a large amount of heat to develop during the threading operation. To reduce the heat, the operator should use a flood of coolant, oil, or tapping paste.

Unless specified otherwise, all threads are right-hand threads. *Right-hand threads* advance clockwise, while *left-hand threads* advance counterclockwise. All threads are single-lead unless specified otherwise. *Single-lead threads* advance one turn for one pitch. *Double-lead threads* advance two pitches in one turn. In practice, this thread is known as a *two-start thread.*

When programming the threading operation, the programmer must consider the minimum distances for acceleration and deceleration of the cutting tool (Figure 7–10). The **acceleration** distance, specified as the L1 dimension, is the minimum distance required for initial running before the cutting tool reaches the programmed speed from the stop-condition. **Deceleration,** specified as the L2 dimension, is the minimum distance required for the cutting tool to stop the motion.

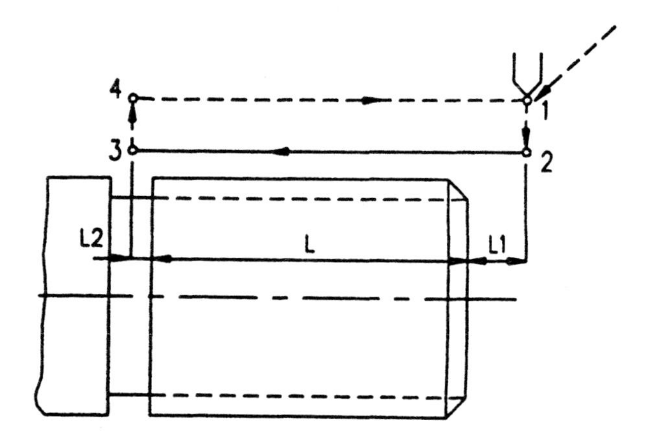

Figure 7–10 Acceleration and deceleration of the threading tool.

The L1 and L2 values depend on the pitch of the lead screw of each particular machine. For the majority of lathes, these values may be calculated by the following formulas where RPM is spindle revolution per minute, and P is the thread pitch.

$$L1 = \frac{RPM \bullet P}{400} \qquad L2 = \frac{RPM \bullet P}{2000}$$

Following are the calculations for acceleration and deceleration for ⅛ thread pitch and 400 spindle RPM:

$$L1 = 400 \bullet 0.125 / 400 = 0.125 \text{ inch}$$
$$L2 = 400 \bullet 0.125 / 2000 = 0.025 \text{ inch}$$

In practice, the programmer rounds off these values to the higher values. For example, L1 = 0.2 inch and L2 = 0.05 inch.

Figure 7–10 shows the tool movements in a threading operation, creating a rectangle in space following this pattern:

From point 1 to point 2, the tool moves in rapid on the X axis to take a threading pass according to the depth of cut.

From point 2 to point 3, the tool feeds on the Z axis, cutting the thread to the programmed length.

From point 3 to point 4, the tool feeds out from the part on the X axis.

From point 4 to point 1, the tool returns in rapid to the cycle start position.

When cutting a taper thread, the tool moves from point 2 to point 3 on the X and Z axes simultaneously. The control repeats these motions until the **thread root diameter** is reached. Programmers sometimes repeat the last pass (called a **spring pass**) to get a better finish on the flanks. (Avoid a spring pass when threading on stainless steel; it may harden the thread. Then, if the thread is not cut to size, the insert breaks when re-threading the part.)

The depth of threading passes depends on the thread form, thread height, and rigidity of the setup. In general, the first cut is deeper (usually from 0.010 inch to 0.030 inch), then the cutting depth should decrease (usually from 0.001 inch to 0.006 inch) because the heat is developing on the sides of the thread.

Programming Threading in a Single Block by the G76 Cycle

On the majority of lathes, threading is programmed through the use of the G76, G92, and G32 canned cycles. The G76 canned cycle allows threading on the outside or inside diameter in a single line of information. This cycle is used the most since it is the easiest and fastest way of programming the threading operation. Following is the instruction format:

G76 X__ Z__ I__ K__ D__ A__ F__;

Address Description:

X The X coordinate of the thread root diameter.
Z The Z coordinate of the thread length.
I The taper change from the start to the finish point of the tool travel (radius value). If the programmer omits or enters I0, a straight thread is cut.
K Single depth of thread or the height of thread (radius value).
D The depth of the first pass (radius value); decimal point format cannot be used.
A An included angle of the threading tool, or thread included angle.
F Thread pitch or lead.

To achieve a more accurate thread pitch or lead, on a majority of controls the programmer can use the address E (with six decimals) instead of the address F (with four decimals). For example, for a 13-thread-per-inch thread, the programmer enters E0.076923 ($\frac{1}{13}$ = 0.076923). Using the address F, he or she can express the pitch or lead with only four decimals, F0.0769. Programmers prefer using the address F, since in most cases the loss of absolute accuracy will not affect the quality of the thread.

Most controls perform automatic chamfering out on the end of thread; thus, it does not have to be programmed. When the programmer does not want chamfering out, he or she enters the appropriate M code on the threading line. For this purpose, some controls use the M24 code; others use the M91 code. Following is a programming example for the part shown in Figure 7–11. The thread is 1.0 UN-8 TPI, and the material is mild steel.

In order to enter proper information in the G76 block, the programmer has to calculate the thread data. Then, a complete program may be designed as follows.

Thread Data:

Pitch: P = 1/Number of threads per inch = $\frac{1}{8}$ = 0.125
Single depth: S = 0.61343 • P = 0.0766
Root diameter: D – (2 • S) = 1.0 – (2 • 0.0766) = 0.8467

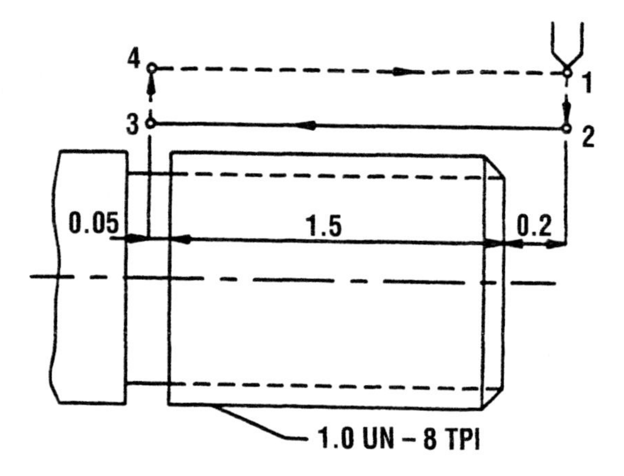

Figure 7–11 Programming a straight thread by the G76 cycle.

When calculating the thread root diameter, the programmer subtracts a double-thread height from the outside diameter for external threading or adds the double-thread height to the bore diameter for internal threading.

Program:

```
G50 X10.0 Z2.0 M42; (Coordinate system preset. High range.)
G00 T0500; (Indexing the tool.)
G97 S650 M03; (Constant RPM. Spindle start.)
G00 X1.2 Z0.25 T0505 M08; (Rapid to position. Offset call.)
G76 X0.8467 Z-1.55 K0.0766 D100 A60 F0.125; (Threading cycle is specified.)
G00 X10.0 Z2.0 T0500 M09; (Cycle cancel. Return home. Offset cancel.)
M01; (Optional program stop.)
```

The tool is brought in rapid to the cycle start point (X1.2 Z0.25); in general, this should be the highest tool X position when cutting an external thread, or the lowest tool X position when cutting an internal thread. Then the threading cycle is specified as follows:

X0.8467 is the X coordinate of the thread root diameter.
Z-1.55 is the Z coordinate of the tool travel, including deceleration amount.
K0.0766 is single depth of thread or the height of thread.
D100 is the depth of the first pass (0.010 inch). Programmers who want the thread to be machined in fewer passes can enter a larger value.
A60 is a 60-degree thread angle.
F0.125 is the thread pitch.

Following is a programming example to cut a taper thread for the part shown in Figure 7–12. The thread is 3.0 UN-4 TPI, and the material is mild steel.

Thread Data:

Pitch: $P = 1/4 = 0.25$
Single depth: $S = 0.61343 \cdot P = 0.61343 \cdot 0.25 = 0.1533$
Root diameter: $D - (2 \cdot S) = 3.0 - (2 \cdot 0.1533) = 2.694$

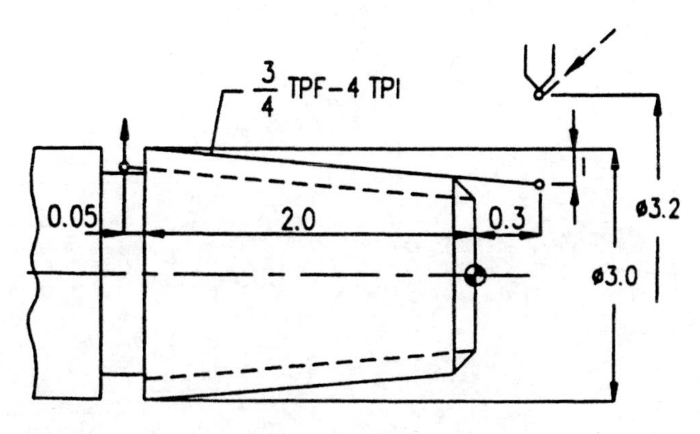

Figure 7–12 Cutting the taper thread by the G76 cycle.

Taper change on radius (the I value) is calculated as follows:

The taper given as taper-per-foot data has to be converted to taper-per-inch data, as is the other data. Thus,

$$0.75 / 12 = 0.0625$$

This is the amount of taper change on diameter per one inch of taper length. The length of the tool travel, including the acceleration and deceleration, is calculated as:

$$L = 2.0 + 0.3 + 0.05 = 2.35$$

This value is multiplied by the amount of taper change on diameter. Thus,

$$2.35 \bullet 0.0625 = 0.1468$$

This is the change in part diameter on the length of tool travel. To find the change on radius, this amount is divided by 2, as follows:

$$0.1468 / 2 = 0.0734$$

This is the amount of taper change on radius for the *total tool travel* programmed using the I address.

The I value may be programmed using a negative or positive sign, depending on whether the thread to be cut is external or internal (Figure 7–13).

Program:

```
G50 X10.0 Z2.0 M42; (Coordinate system preset. High range.)
G00 T0400; (Indexing the tool.)
G97 S410 M03; (Constant RPM. Spindle start.)
G00 X3.2 Z0.3 T0404 M08; (Rapid to position. Offset call. Coolant on.)
G76 X2.694 Z-2.05 I-0.0734 K0.1533 D150 A60 F0.25; (Threading cycle.)
G00 X10.0 Z2.0 T0400 M09; (Cycle cancel. Return to origin. Offset cancel.)
M1; (Optional program stop.)
```

The control moves the tool for the first pass in a plus or minus direction on the X axis, depending on the sign specified by the address I. When cutting the external thread, the programmer enters a minus sign; when cutting the internal thread, he or she enters a plus sign. On machines with a left-hand coordinate system, the sign for the I address is opposite: a minus sign is used for internal threading, and a plus

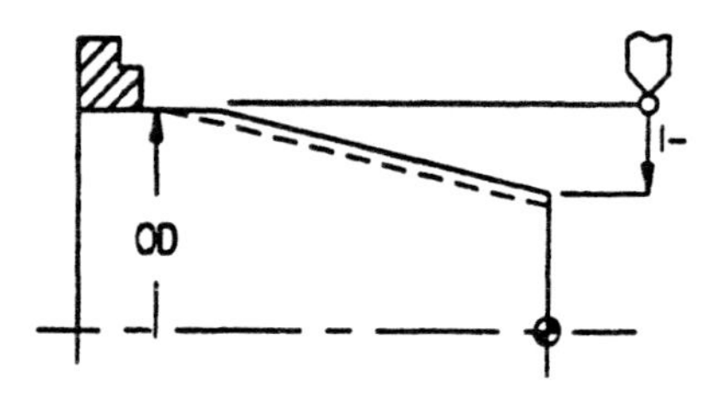

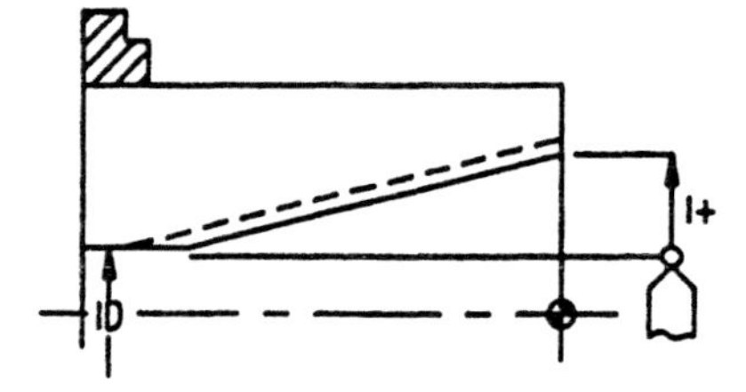

Figure 7–13 A sign for the I address.

sign is used for external threading. When calculating the I value, he or she must include the acceleration and deceleration in total tool travel.

On some of the latest controls, the G76 cycle is programmed in two blocks:

First block:

G76 P021060 Q20 R30;

Address Description:

P 02	Number of repeats of the last pass (Two repeats are programmed.)
10	Chamfer amount (One full chamfer is programmed.)
60	Thread relief in degrees (A 60-degree thread angle is to be cut.)
Q 20	Minimum depth of cut (A 0.002 minimum depth of cut is programmed.)
R 30	Finishing allowance before the last pass (A 0.003 inch is programmed.)

Second block:

G76 X3.27 Z-2.6 P650 Q150 R-0.0843 F0.125;

Address Description:

X	Thread root diameter.
Z	Thread length.
P	Single depth of thread; decimal point format cannot be used.
Q	Depth of the first pass; decimal point format cannot be used.
R	Taper amount on radius (the same as the I value).
F	Thread pitch.

Designed in this way, the G76 canned cycle has one distinct advantage: It enables the programmer to control the threading parameters without depending on the machine parameter setting.

The G32, the Most Versatile Threading Cycle

The G32 canned cycle is the most versatile threading cycle, allowing the cutting of straight, taper, and scroll screws, as well as tapping, on the lathe. The G32 cycle allows the programmer to affect the depth of each pass and repeat any pass as he or she wishes; this is the advantage over the G76 cycle. When combined with a subprogram, it is easy and fast to program this cycle. The instruction format is:

G32 Z__ F__; (Straight screw.)
G32 Z__ I__ F__; (Taper screw.)
G32 X__ F__; (Scroll screw when face threading.)

Address Description:

X	The X coordinate of the thread root diameter.
Z	The Z coordinate of the thread length.

I The taper change from the start to the finish point of the tool travel (radius value). If the programmer omits or enters I0, a straight thread is cut.

F Feed rate for threading; equal to the lead or pitch for a single start thread.

Following is a programming example to cut a straight screw for the part shown in Figure 7–14. The thread is 1.25 UN-10 TPI. The material is stainless steel.

Thread Data:

Thread pitch: P = 1⁄10 = 0.1
Thread depth: S = 0.61343 • P = 0.61343 • 0.1 = 0.0613
Thread root diameter: D – (2 • S) = 1.250 – (2 • 0.0613) = 1.1274

Program:

```
G50 X15.0 Z3.0 M42; (Coordinate system preset; point A.)
G00 T0303; (Tool and tool offset call.)
G97 S475 M03; (Spindle start.)
G00 X1.5 Z0.3 M08; (Rapid to position; point B.)
X1.230; (Positioning for the first pass; point C.)
M98 P50; (Subprogram call.)
G00 X1.220; (Second pass.)
M98 P50; (Subprogram call.)
G00 X1.210; (Third pass.)
M98 P50; (Subprogram call.)
...; (The program continues; the depth decreases.)
X1.1274; (Last pass.)
G00 X15.0 Z3.0 T0300 M09; (Return home. Offset cancel. Coolant stop.)
M01; (Optional program stop.)

O50; (Subprogram number.)
```

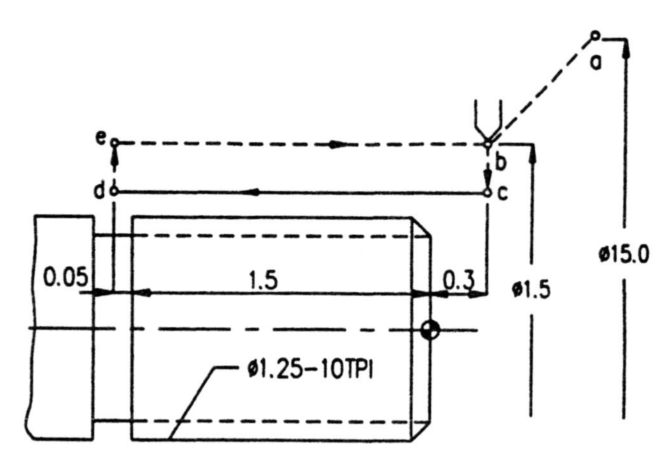

Figure 7–14 Cutting a straight screw by the G32 cycle.

G32 Z-1.55 F0.1000; (Threading up to point D.)
G00 X1.5; (Rapid out; point E.)
Z0.3; (Return to start position; point B.)
M99; (Return to the main program.)

For each pass the tool is positioned in the main program. The subprogram repeats the rest of the tool movements. If the programmer does not use the subprogram, he or she must repeat all of the tool movements. This will result in a longer program.

When tapping on the lathe, programmers use the G32 cycle. After the hole is drilled, the spindle is turned clockwise and the tap is moved in rapid to the part centerline. Then the programmer calls the G32 cycle and tapping in starts. When the required depth is reached, the spindle stops and turns in a counterclockwise direction for tapping out. The programmer must use the same RPM and feed rate when tapping in and out of the hole. Following is a programming example for ½-13 UNC tap, ¾ inch deep.

G50 X13.0 3.0 M41; (Coordinate system preset. Low range.)
G00 T0505; (Tool and tool offset call.)
G97 S125 M03; (Spindle start in the clockwise direction.)
G00 X0 Z0.1 M08; (The tap is at the part centerline.)
G32 Z-0.95 F0.0769; (Tapping in.)
G00 M05; (Cycle cancel. Spindle stop.)
G32 Z0.1 F0.0769 S125 M04; (Tapping out, spindle start counterclockwise.)
G00 X13.0 Z3.0 T0500 M09; (Return home. Offset cancel. Coolant stop.)
M01; (Optional program stop.)

For tapping on the lathe, the operator should choose a floating tap holder in order to amortize the tool travel until the spindle is stopped completely. When the rigid tap holder is used, the tap breaks.

The G92 Cycle, the Accelerated G32 Cycle

The G92 cycle is an accelerated G32 cycle that works on the same principle. For this cycle, the programmer need only specify the X coordinate for each threading pass. The control returns the tool to the start point after the programmed thread length is reached. The G92 cycle allows cutting both straight and taper screws, either on the outside or the inside diameter. Following is the instruction format:

G92 X__ Z__ F__; (Straight screw.)
G92 X__ Z__ I__ F__; (Taper screw.)

Following is a programming example of threading a straight screw for the part shown in Figure 7–15. The thread is 1.25 UN-10 TPI and the material is mild steel.

Thread Data:

Pitch: P = 1⁄10 = 0.1
Thread depth: S = P • 0.61343 = 0.1 • 0.1343 = 0.0613
Thread root diameter, D – (2 • S) = 1.250 – (2 • 0.0613) = 1.1274

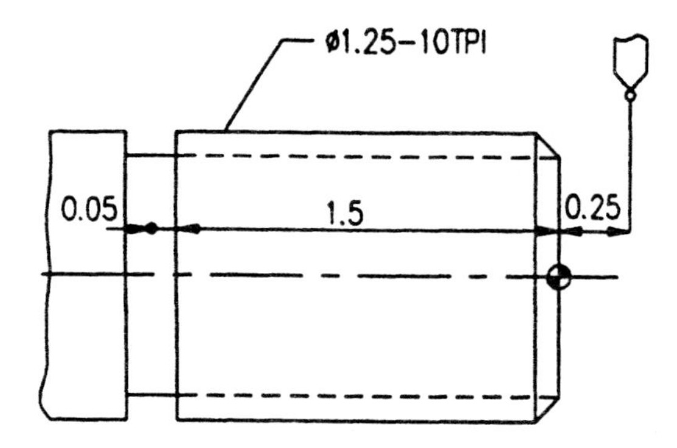

Figure 7–15 Cutting a straight screw by the G92 cycle.

Program:

```
G50 X10.0 Z2.0 M42; (Coordinate system preset. High range.)
G00 T0505; (Tool and tool offset call.)
G97 S520 M03; (Spindle start; constant RPM.)
G00 X1.35 Z0.25 M08; (Rapid to position. Coolant on.)
G92 X1.23 Z-1.55 F0.1; (First pass.)
X1.22; (Second pass.)
X1.21; (Third pass.)
...; (The program continues; the depth decreases.)
X1.1274; (Last pass.)
X1.1274; (Last pass repeated.)
G00 X10.0 Z2.0 T0500 M09; (Cycle cancel. Return home. Offset cancel.)
M01; (Optional program stop.)
```

The Z coordinate for the thread length is programmed for the first pass; for the rest of the passes it does not have to be programmed, since the control uses similar subprogramming technique as presented in this text for the G32 cycle. As true for the G32 cycle, the programmer can affect each threading pass by choosing a different depth or by repeating the pass.

Cutting Deep or Wide Threads

In a threading operation the tool sometimes chatters, especially when machining deep or wide threads. The chatter must be eliminated or minimized, or the machined thread will not seal properly. Following are the main reasons for tool chatter:

The part is not supported.
The tool overhangs too much.
The cutting speed is incorrect.
The depth of the first pass is incorrect.

When threading, the tool has a tendency to push or move the part. To prevent this, the operator should support the part by pushing a short part against the stop in the chuck and using the stop and tailstock center to support a long part. He or she should also shorten the tool overhang as much as possible; otherwise, both the threading bar and the external threading tool may produce the chatter.

The programmer must choose a proper cutting speed for a particular material. When needed, the operator should adjust the speed according to cutting conditions; sometimes even a proven cutting speed does not give the expected result when the same job is run again, because the material may be of a different hardness, the tool overhang may be different, or the operator might be using a different type of insert.

The depth of the first pass is important for both tool life and thread finish. The first threading pass should not be too excessive, but it also should never be too shallow. In both cases the first pass may produce chatter that may cause the stress of material. Then the chatter stays through the rest of the passes.

Even when the programmer and the operator take all of the precautionary steps, the tool still may chatter when cutting deep or wide threads. To solve this problem, the programmer can reprogram the threading operation for each flank of the thread, cutting only a minimum of material. Following is an example for a 0.16-inch-deep external thread on a 1.75-inch-long screw with a pattern of 4 threads per inch on a 5.00 inch diameter.

```
G00 X5.2 Z0.5; (Rapid to position for threading.)
G76 X4.68 Z-1.75 K0.16 D300 F0.25; (Threading cycle is specified. The depth
    of the first pass is 0.03 inch.)
G00 X5.2 Z0.503; (Tool start point moved away from the part for 0.003 inch.)
G76 X4.68 Z-1.75 K0.16 D1200 F0.25; (Repeating the threading operation
    using a 0.12 inch depth of cut; cutting the flank on the right-hand side.)
G00 X5.2 Z0.497; (Tool start point moved closer to the part for 0.003 inch.)
G76 X4.68 Z-1.75 K0.16 D1200 F0.25; (Repeating the threading operation
    using a 0.12 inch depth of cut; cutting the flank on the left-hand side.)
G00...; (Return home.)
```

In this program, the first G76 line cuts the thread in the ordinary manner using a 0.030 inch depth of cut. Then the tool start position moves for a 0.003 inch on the Z axis; first away and then toward the part. Using a 0.12 inch depth of cut, the G76 cycle repeats twice to clean up the flanks on each side. This finishes in a few passes since the depth of the first pass is about 75 percent of the full threading depth. On some controls, the programmer can use special parameters when cutting deep or wide threads by the G76 cycle. These parameters (P1, P2, P3, and P4) allow cutting the thread with constant depth of each pass or with constant amount of material to be removed in each pass.

Cutting Left-Hand Threads

The programmer can program cutting a left-hand thread using either the M04 code or the M03 code. The M04 code turns the spindle in a counterclockwise direction; then the tool start position is at the beginning of the thread, and on most lathes the tool must be set up with the tip upward [Figure 7–16, position (a)].

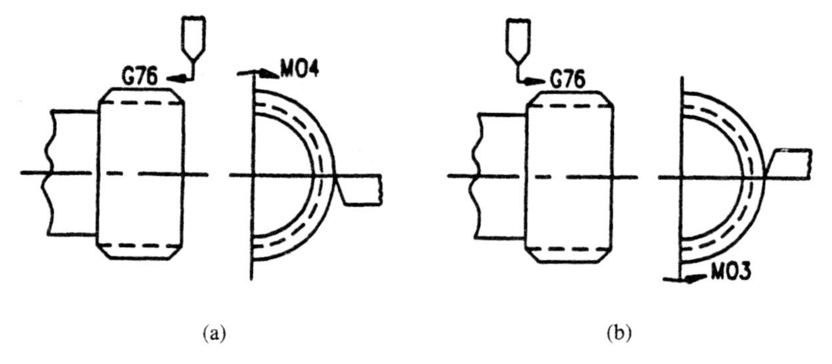

Figure 7–16 Cutting a left-hand thread.

The M03 code turns the spindle clockwise; then the tool start position is at the end of the thread and on most machines the tool must be set up with the tip downward (position b). Both of these techniques are presented in the following series of instructions to cut a 1.00-inch-diameter thread, 10 threads per inch, 0.06 inch deep, 1.25 inches long.

Using the M04 Code:

G97 S430 M04; (Spindle counterclockwise.)
G00 X1.15 Z0.2; (Rapid to the start position at the beginning of the thread.)
G76 X0.88 Z-1.3 K0.060 D100 F0.1 A60; (Threading cycle is specified.)

Using the M03 Code:

G97 S430 M03; (Spindle clockwise.)
G00 X1.15 Z-1.3 M08; (Rapid to the start position at the end of the thread.)
G76 X0.88 Z0.2 K0.060 D100 F0.1 A60; (Threading cycle is specified.)

These two programming segments differ in two things: the use of the M03/M04 code, and tool start and finish point on the Z axis. The operator must also distinguish them by setting up the threading tool accordingly.

Cutting Multistart Threads

A **multistart thread** allows fast linear motion per revolution; the more thread starts, the faster linear motion per revolution. On the drawing, the designer may describe a multistart thread through the lead or through the pitch of thread. For example, he or she may indicate 0.125 pitch, three starts, or 0.375 lead, three starts. The programmer must use the lead value, not the pitch value; therefore, for a 0.375 inch lead thread, the feed rate must be programmed as F0.375, not F0.125. When the first start is cut, the distance between two crests is 0.375 inch (a lead value). When all the starts are cut, the distance between two crests is 0.125 inch (a pitch value), because all three starts come beside each other.

When programming cutting a multistart thread, the programmer should keep in mind that the controls differ in this area. All controls allow a programmer to program each thread start independently; he or she can machine the thread for one start and then proceed to the next, and so on for all thread starts. For each start the programmer must move the tool start point on the Z axis for one pitch value, as illustrated in the following instructions to machine a 0.375 thread lead, three starts:

G00 X1.2 Z0.2; (Rapid to the start position.)
G76 X0.88 Z-1.25 K0.06 D0.01 A60 F0.375; (Machining the first start.)
G00 W0.125; (Move over in incremental for one pitch value.)
G76 X0.88 Z-1.25 K0.06 D0.01 A60 F0.375; (Machining the second start.)
G00 W0.125; (Move over in incremental for one pitch value.)
G76 X0.88 Z-1.25 K0.06 D0.01 A60 F0.375; (Machining the third start.)
G00...; (Return home or positioning to another position.)

Some controls are more helpful in programming cutting multistart threads since they allow the programmer to use the address Q to express the number of starts as either a single number or as an angular position.

Example 1:

G00 X1.2 Z0.25; (Rapid to the start position.)
G76 X0.88 Z-1.25 K0.06 D0.01 A60 F0.375 Q3; (Machining of all three starts specified by the address Q.)
G00...; (Cycle cancel; move home or to another position.)

Example 2:

G00 X1.2 Z0.25; (Rapid to the start position.)
G76 X0.88 Z-1.25 K0.06 D0.01 A60 F0.375 Q0; (Machining the first start.)
G76 X0.88 Z-1.25 K0.06 D0.01 A60 F0.375 Q180; (Machining the second start.)
G76 X0.88 Z-1.25 K0.06 D0.01 A60 F0.375 Q270; (Machining the third start.)
G00...; (Return home or positioning to another position.)

In the first example, the control allows the programmer to machine all three starts in one command. In the second example, the programmer must program a separate threading cycle for each thread start; however, he or she does not need to enter the moveover command after each start.

Grooving

When grooving, the programmer and the operator should implement an effective chip control in order to prevent chip buildup under the tool, which damages the tool and/or the part. First, they should select the tools with proper tool geometry for an application. **Tool rake** (which ranges from positive to negative, producing different kinds of chips) is an important aspect of tool geometry. Tools with a

zero and negative rake are stronger and more durable. However, **positive rake** produces curled chips, which are desired when grooving. Next, chatter and unsatisfactory finishing are constant problems when grooving. Some of the following tips may be helpful for avoiding chatter:

Lower the cutting speed using 40–60 percent of RPM for turning.
Shorten the overhang of the grooving tool to lower vibration.
Decrease the feed rate using 40–60 percent of the feed rate for turning.
Increase the feed rate when grooving stainless steel.
On a tool with zero rake, grind a positive rake to get curled chips.
Use dwell when the tool reaches the groove bottom to get a better finish.
Use constant surface speed control to improve the finish on the groove sides.

The finish at the groove bottom can be improved by using the G01 function and single cuts to break the chips before the final groove diameter. The programmer should program the tool to pull out from the groove when the groove is close to size. The coolant will splash the part and insert, lowering the heat. Then the programmer should increase the cutting speed and program the final groove diameter. Finishing the groove bottom may also be improved if the dwell is programmed when the tool reaches the final depth of groove. This is especially true when grooving a large diameter, which is normally programmed to rotate slowly. The dwell allows a full turn of the part before the tool pulls out. This results not only in a better surface finish, but also in a groove diameter that is cylindrical.

Diameter Grooving by the G75 Cycle

Programmers use the G75 cycle for grooving on the part outside or inside diameter. This cycle is a **peck grooving cycle** along the X axis, meaning that the tool pecks into material by the programmed feed, then stops and retracts by the amount specified by the parameter. At the same time, chips are broken, which is important when deep grooving. The tool continues peck-in and peck-out until the groove depth is reached. Then the tool returns to its start position. If the groove is wide and more than one pass is needed, the control moves the tool to another X position and the process repeats. The same is true when several grooves are to be cut. After the cycle is finished, the tool returns to the cycle start point. Following is the instruction format for the G75 cycle:

G75 X__ Z__ I__ K__ D__ F__;

Address Description:

X The final depth of groove on the X axis.
Z The position of the last groove on the Z axis, or the tool final Z position when cutting a wide groove.
I The depth of cut for each peck on the X axis (radius value). The sign is not needed.

K The distance between the grooves on the Z axis, or moveover distance when cutting a wide groove.
F The feed rate.

When a groove wider than the tool width is to be cut, the value for the address K should be less than the tool width in order to achieve the overlapping.

In order for the operator to set up the tool and machine the part properly, the programmer should make clear which point on the tool tip is to be used as the reference point. Following is a programming example to cut a wide groove on the part shown in Figure 7–17. The material is 4140 steel.

```
G50 X15.0 Z3.0 M41; (Coordinate system preset. Slow range.)
G00 T0404; (Tool and tool offset.)
G97 S215 M03; (Spindle start.)
G00 X6.1 Z0 M08; (Rapid to position.)
Z-0.7 M08; (Positioning on the Z axis: 0.5 + 0.2 tool width = 0.7.)
G75 X5.0 Z-1.3 I0.05 K0.17 F0.004; (Diameter grooving cycle.)
G00 X15.0 Z3.0 T0400 M09; (Cycle cancel. Return home. Offset cancel.)
M01; (Optional program stop.)
```

In this program, the K value is 0.17, which is less than the tool width of 0.2 inch. Thus, overlapping is achieved. The programmer can start grooving on the farther groove end, and he or she can use the back end of the tool as the reference point. Following is an example for cutting three identical grooves on the part shown in Figure 7–18. The material is mild steel.

```
G50 X15.0 3.0 M41; (Coordinate system preset. Low range.)
G00 T0303; (Tool and tool offset call.)
G97 S225 M3; (Spindle start.)
G00 X6.1 Z-0.3; (Rapid to position.)
G75 X5.0 Z-1.3 I0.05 K0.5 F0.005; (Diameter grooving cycle.)
G00 X15.0 Z3.0 T0300 M09; (Cycle end. Return home. Offset cancel.)
M01; (Optional program stop.)
```

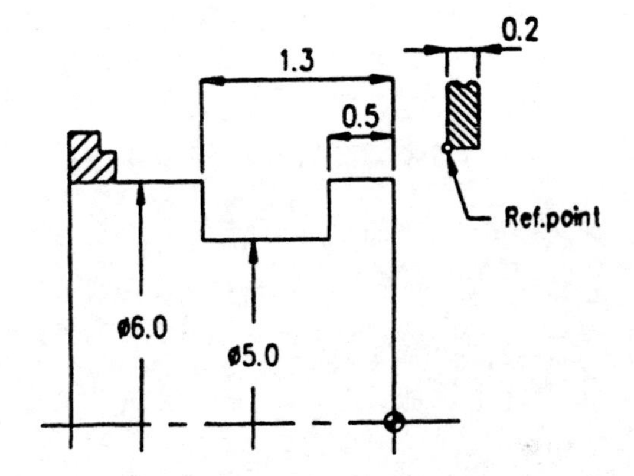

Figure 7–17 Cutting a wide groove by the G75 cycle.

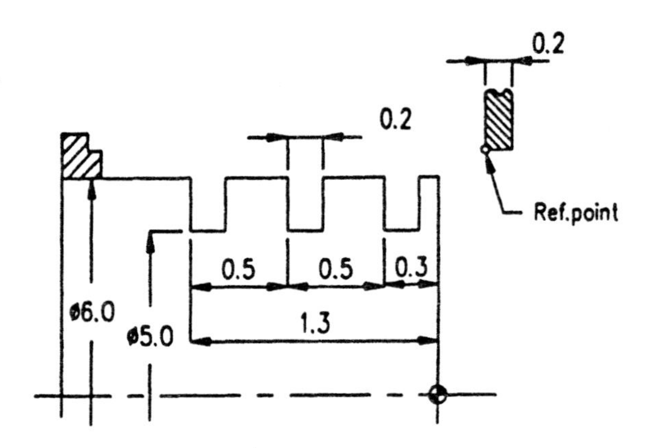

Figure 7–18 Cutting three grooves on the part diameter.

The position of the first groove is specified as the start point of the G75 cycle (Z-0.3). The position of the last groove is controlled by the instruction Z-1.3. The distance between the grooves is programmed by the word K0.5.

Face Grooving by the G74 Cycle

Programmers use the G74 cycle for grooving on the part face. This cycle is a peck grooving cycle along the Z axis; the tool pecks into material by the programmed feed, then stops and retracts by the amount specified by the parameter. At the same time, chips are broken, which is important when deep grooving. The tool continues peck-in and peck-out until the groove depth is reached; then the tool returns to its start position. If the groove is wide and more than one pass is needed, the control moves the tool to another X position and the process repeats. The same is true when several grooves are to be cut. After the cycle is finished, the tool returns to the cycle start point. Programmers also use this cycle for peck drilling, which is a good way of deep hole drilling when ordinary drilling methods result in chipping problems. Following is the instruction format for the G74 cycle:

G74 X__ Z__ I__ K__ D__ F__;

Address Description:

- X The position of the last groove on the X axis, or final tool X position when cutting a wide groove.
- Z The final depth of groove on the Z axis, or drilling depth.
- I The distance between the grooves on the X axis, or moveover distance when cutting a wide groove (radius value). The sign is not needed.
- K The depth of cut for each peck on the Z axis.
- F The feed rate.

Following is an example for cutting a wide groove on the part shown in Figure 7–19. The material is mild steel.

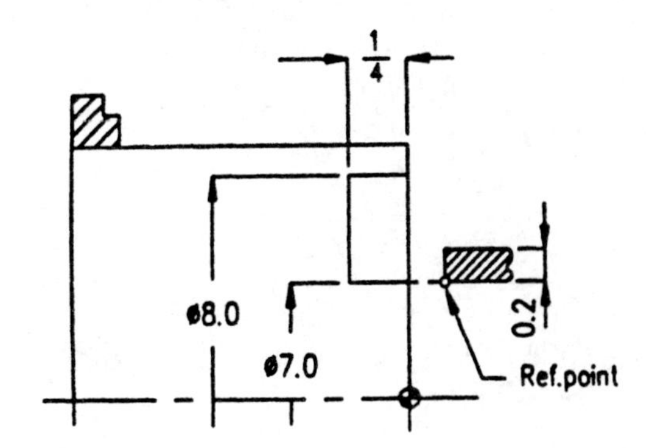

Figure 7–19 Cutting a wide groove by the G74 cycle.

```
G50 X15.0 Z3.0 M41; (Coordinate system preset.)
G00 T0303; (Tool and tool offset call.)
G96 S215 M03; (Spindle start.)
G00 X7.0 Z0.5 M08; (Rapid to position.)
Z0.05; (Closer to the part on the Z axis.)
G74 X7.6 Z-0.25 I0.17 K0.05 F0.004; (Face grooving cycle.)
G00 X15.0 Z3.0 T0300 M09; (Cycle end. Return home. Offset cancel.)
M01; (Optional program stop.)
```

The tool reference point is on the lower corner; the cutting starts at 7.0 inch diameter and finishes at 7.6 inch diameter because the double tool width is subtracted from 8.0 inch diameter. The cutting may be started at the bigger part diameter. The start-up block would be:

```
G00 7.6 Z0.5;
Z0.05.1;
G74 X7.0 Z-0.25 I0.17 K0.05 F0.004;
```

The programmer can use the G74 cycle to cut a series of identical, equally spaced grooves, as shown in the following program for the part in Figure 7–20.

```
G50 X15.0 Z3.0 M41; (Coordinate system preset.)
G00 T0303; (Tool and tool offset call.)
G96 S250 M03; (Spindle start.)
```

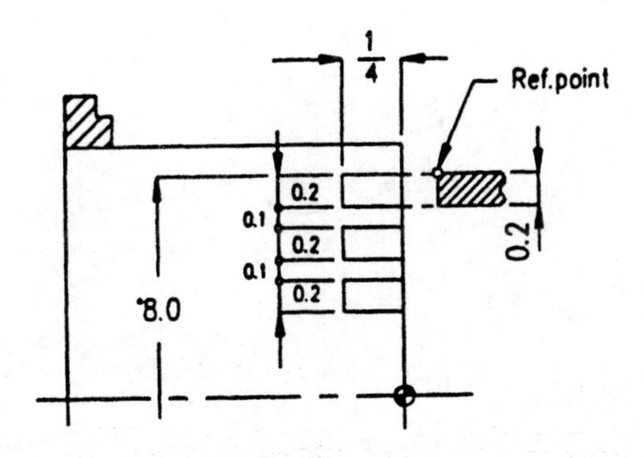

Figure 7–20 Cutting a series of grooves on the part face.

```
G00 X8.0 Z0.5; (Rapid to position.)
Z0.1; (Closer to the part on the Z axis.)
G74 X6.8 Z-0.25 I0.3 K0.05 F0.005; (Face grooving cycle.)
G00 X15.0 Z3.0 T0300 M09; (Cycle end. Return home. Offset cancel.)
M01; (Optional program stop.)
```

The control uses the I address to shift the tool to the next groove position and to keep the same distance between the grooves.

Summary

CNC machine builders have designed built-in programs for the machine's electronic circuits. These are known as canned cycles. There are canned cycles for face and diameter roughing, finishing, profile copying, grooving on face and diameter, and threading. Each canned cycle has a set of rules that the programmer should follow by entering such constants as depth of cut, finishing allowance, drilling depth, thread pitch, and so on.

Lathe canned cycles are divided in two groups: fixed cycles and multiple repetitive cycles. Fixed cycles permit programming three or four successive tool movements in a single block. They also permit repetition of the operations with a small change in the next block. The following are canned cycles:

G32	Threading cycle
G90	Diameter cutting cycle
G92	Threading cycle
G94	Face cutting cycle

Note that the G90 and G94 fixed cycles are not currently used much, since more advanced canned cycles have replaced them.

Multiple repetitive cycles are also known as automatic repeat cycles. They are more sophisticated than fixed cycles. In automatic repeat cycles, the tool cuts the material in repeating sequences until the specified profile is achieved. There is no limit to the number of repeat passes. The control calculates the passes according to specified data. Each of these cycles has its particular purposes, but they are all similar to program. On the CNC lathe, the following repetitive canned cycles are commonly used:

G71	Diameter roughing cycle
G72	Face roughing cycle
G73	Pattern repeat cycle
G70	Finishing cycle
G74	Face peck-grooving cycle
G75	Diameter peck-grooving cycle
G76	Threading cycle

Programmers use the G71 canned cycle for roughing parts along the Z axis. They also use the G72 cycle for facing and roughing parts that are larger in diameter than in length; the tool cuts the part along the X axis. When using the G71 or G72 roughing cycle, the programmer needs to specify the finish part profile between the P and Q blocks. The control calculates the number of passes, depending on the amount of material to be removed. After roughing is completed, the tool leaves a specified amount of material for a finishing cycle that can be removed by the same or a different tool.

The feed rate specified in the roughing cycle will be in effect for all of the roughing passes. When the programmer wants a variation in surface finishing, he or she must program the feed rate and spindle speed inside the roughing cycle. They will be in effect if the finishing cycle is programmed after the roughing cycle. When the programmer does not want the finishing cycle, he or she omits or specifies zero values for U and W in the roughing cycle. The tool will then cut the part profile to a finish size in a roughing cycle.

The G71 canned cycle has four different patterns. In each of them, the part diameter successively increases or decreases. The programmer can use automatic tool radius compensation in the multiple repetitive cycles. To enable the control to pick up the compensation in the finishing cycle, he or she must program the compensation call code inside the roughing cycle.

When roughing parts with a profile already preformed by casting or forging, it is not efficient to use the G71 and G72 cycles. Time loss occurs when the tool makes a number of unnecessary passes cutting air. Machining time is the same as if the part was not preformed. To avoid this, programmers use the G73 cycle with copying capabilities. The G73 cycle is derived from the G71 cycle and works on the same principle. The only difference is that the tool does not rough along the Z axis; it follows the preformed profile even when roughing. Consequently, machining time can be saved. When programming the G73 cycle, the programmer specifies the pattern of the finish part profile. The pattern is repeated a number of times (as in the G71 cycle) until the finish part profile is cut. The number of passes depends on the finish part size and the depth of cut.

Programmers program the G70 cycle after the G71, G72, or G73 multiple repetitive roughing cycles to cut the part profile to a finish size; the tool cuts the part in one pass, following the part profile specified in the roughing cycle. The depth of cut for the finishing pass depends on the finishing allowance programmed by the U and W addresses in the roughing cycle. Any variations in the feed rate and cutting speed programmed in the roughing cycle come into effect when the finishing cycle is executed. The block numbers designated by the P and Q addresses are the same block numbers programmed in the roughing cycle. This means that the programmer cannot affect the part profile through the use of the G70 cycle. If he or she wants any alteration in the tool path, feed rate, or cutting speed, he or she must use the roughing cycle. The P and Q block numbers must be in the memory of CNC machine. If the control cannot find them, it will display an alarm.

A programmer can call the finishing cycle immediately after the roughing cycle is finished using the same tool. The end point in roughing then becomes the start point in finishing. Also, he or she can call the finishing cycle at any point later in the program, using the same or a different tool for roughing. Usually, finish with a different tool is preferred so the size is held better.

Thread-cutting operations are performed when the CNC machine is in a special state that limits operator input during program execution. Therefore, special caution is needed when programming, as well as when setting up the machine for threading operations. The operator cannot adequately stop the machine once it starts threading since the Feed Hold function usually has no effect. Single-block execution *is* possible because the cutting tool stops after it returns to the cycle start point. If it appears that the tool may collide with the part or chuck, the operator should use the EMERGENCY STOP button to stop the machine.

Threading feed rates depend on the thread pitch. They are much higher than turning or boring feed rates. This causes a large amount of heat to develop when threading. To reduce the heat, the operator should use a flood of coolant, cutting oil, or cutting paste.

Unless specified otherwise, all threads are right-hand threads. Right-hand threads advance clockwise, while left-hand threads advance counterclockwise. The number of turns of the thread and the feed rates are obtained from the lead of the thread. All threads are single-lead unless specified otherwise. Single-lead threads advance one pitch for one turn. Double-lead threads advance two pitches in one turn; in practice, this thread is known as a two-start thread.

When programming the threading operation, the programmer should care for the minimum distances for the acceleration and deceleration of the cutting tool. The acceleration is the minimum distance required for initial running before the cutting tool reaches the programmed speed from the stop condition. The deceleration is the minimum distance required for the cutting tool to stop motion.

The G76 canned cycle allows threading on the outside or inside diameter in a single line of information. Programmers use this cycle the most since it is the easiest and fastest way to program the threading operation. He or she needs to specify constants such as the X coordinate of the thread root diameter and the height of thread. The control then calculates the number of passes needed to finish the thread.

The G32 cycle is the most versatile threading cycle, allowing the programmer to cut straight, taper, and scroll screws, as well as to perform tapping, on the lathe. When using the G32 cycle for tapping on the lathe, the programmer uses the M03 code when tapping in and the M04 code when tapping out to form the hole.

The G92 cycle is the accelerated G32 cycle, which works on the same principle. For this cycle the programmer must specify the X coordinate for each pass. When the programmed thread length is reached, the tool pulls out from the thread and returns to the cycle start point. Then the tool moves to the position of the next pass.

To machine a left-hand thread, the programmer can choose to turn the spindle counterclockwise. Then the tool start point on the Z axis will be at the front of the part, as when cutting right-hand thread. He or she can also choose to turn the spindle clockwise. Then the tool start point will be at the end of the thread.

On any control the programmer can program a multistart thread by repeating a threading operation a number of times, depending on the number of the thread starts. He or she would then move the tool start position for one pitch value and repeat the threading cycle for the next thread start. Some controls use the Q address to help program a multistart thread.

Finishing of the groove bottom may be improved if a dwell is programmed. This is especially true when grooving a large diameter, which is normally programmed to rotate slowly. The dwell allows a full turn of the part before the tool pulls out. This results not only in a better surface finish, but also in a groove diameter that is cylindrical.

Programmers use the G75 cycle to cut one or several grooves on the outside or inside part diameter. A groove of the same width as the tool width is usually cut, but a wider groove than the tool width may also be programmed. This cycle is known as the peck grooving cycle, since the tool pecks into material by the specified feed rate, then stops cutting and retracts for the amount specified by the parameter. This process is repeated until the depth of groove is reached. When the groove is wider than the tool width, or several grooves on the same diameter are to be cut, the tool is moved to the next Z position automatically. After the cycle is finished, the tool returns to the cycle start point.

The G74 cycle is used for face grooving. It is a peck grooving cycle along the Z axis. If the groove is wide and more than one pass is needed, the tool moves to another X position and repeats the process. The same is true when several grooves are to be cut. This cycle may also be used for peck drilling, which is a good way of deep hole drilling when ordinary drilling methods result in chipping problems.

Key Terms

acceleration
canned cycles
deceleration
double-lead thread
fixed lathe cycles
left-hand threads
multiple repetitive cycles
multistart thread
peck grooving cycle
positive rake
right-hand threads
single-lead thread
spring pass
thread root diameter
tool rake
zero and negative rake

Self-Test

The answers are in Appendix E.

1. _______________ are the programs built into the machine electronic circuits to simplify programming.
2. _______________ permit programming three or four successive tool movements in a single block.
3. When using _______________, the material is cut in repeating sequences until the specified profile is achieved.
4. When threading, the _______________ of the tool is specified by the minimum distance required for initial running before the cutting tool reaches the programmed speed from the stop condition.
5. In threading, _______________ is specified by the minimum distance required for the cutting tool to stop motion.
6. When calculating _______________, the double-thread depth is added or subtracted from the part diameter.
7. A repeating pass in threading is known as _______________.
8. _______________ advance clockwise.
9. _______________ advance counterclockwise.
10. _______________ advance the thread one pitch for one turn.
11. _______________ advance the thread two pitches for one turn.
12. _______________ may be cut by repeating the threading operation a number of times.
13. _______________ is an important part of tool geometry.
14. The tools with _______________ rake are stronger and more durable.
15. _______________ rake produces curled chips, which are preferred when grooving.
16. In a _______________ cycle, the tool pecks into the material by the specified feed rate, then stops cutting and retracts by the amount specified by the parameter.

Relating the Concepts

No answers are suggested.

1. What types of part profiles can be cut by the G71 cycle?
2. Define the tool start point on the X axis before the G71 cycle is specified:
   When turning.
   When boring.
3. What is the sign for the finishing allowance when boring by the G71 cycle?
4. How would you program variation in surface finish when using multiple repetitive canned cycles?

5. The length of a thread is 1.75 inches. Calculate the I value for a 3/4 taper per foot when threading by the G76 cycle, adding 0.3 inch for acceleration and 0.05 inch for deceleration of the threading tool.
6. A groove is 0.5 inch wide. The tool width is 0.1 inch. Program the K value when using the G75 cycle. Explain.
7. Calculate the root diameter when threading external 2.500 Stub Acme, 8 TPI thread.
8. For the part shown in Figure 7–21,
   a. Select the tools and assign the tool numbers.
   b. Decide which tool to use to set Z0 and where to set it.
   c. Write the program to turn the outside diameter, roughing and finishing.
   d. Write the program to drill the hole through.
   e. Write the program to finish a wide groove using the G75 cycle.
   f. Write the subprogram to cut all 1/8 inch grooves.
   g. Write the program for boring, roughing and finishing.
   h. Write the program to cut the thread undercut using the G01 function.
   i. Write the program to cut the thread using the G76 canned cycle.
   j. Write the program for part-off. Before part-off is completed, remove the sharp edges on the part.

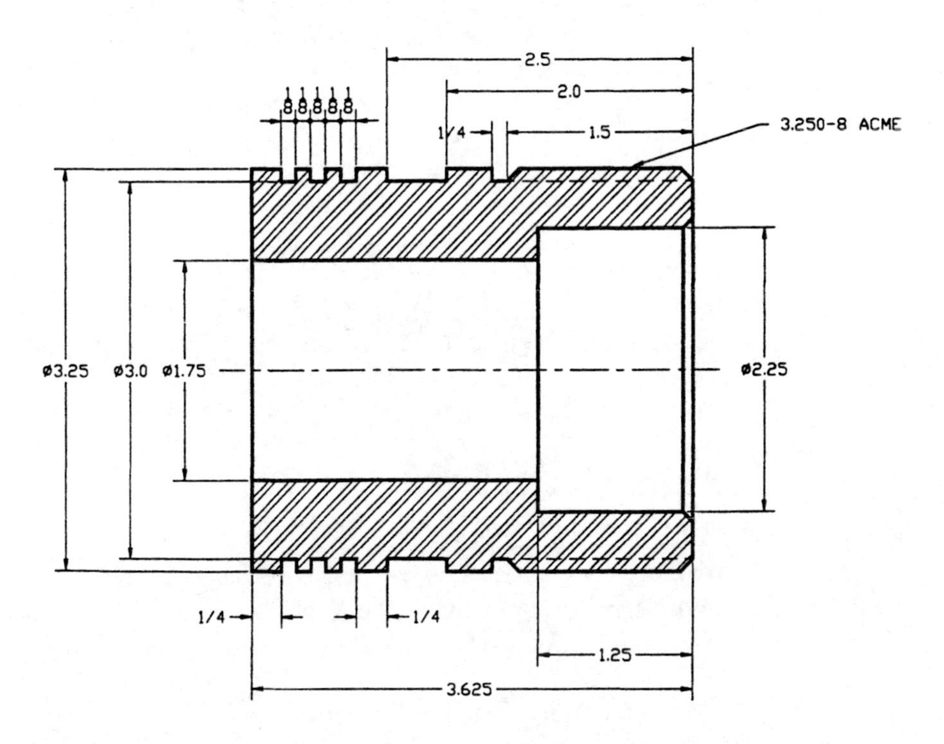

Figure 7–21 Multiple machining sequences.

8 Programming CNC Machining Centers

Key Concepts

Machining Center Canned Cycles

Drilling

- Drilling by the G81 Cycle
- Drilling with Dwell
- Peck Drilling

Tapping

Boring

Machining at Equal Intervals

Milling

Helical Milling

Machining Center Canned Cycles

CNC machining centers (Figure 8–1) perform a number of different operations that can be performed in a single setup, which makes CNC machining centers the most versatile of all machine tools. The most advanced machining centers are known as *contouring machines* capable of up to five-axis simultaneous control. They are equipped with automatic tool changers for a full random tool selection and controls with 32-bit processors.

Figure 8–1 This vertical machining center has a heavy-wall construction that increases cutting rigidity and vibration resistance while enhancing operator accessibility to the fixture and workpiece. (Courtesy Hitachi Seiki U.S.A., Inc.)

When programming for the machining center, remember that the tool, not the table, is programmed to move, although the table brings the part into position for cutting. In fact, the table moves the opposite way from the programmed direction. For example, if the tool is programmed to move 5.0 inches in a positive direction on an axis, the table will move –5.0 inches on the same axis.

Hole machining is the most common operation performed on a CNC machining center. For this reason, canned cycles for hole machining are a standard feature on a majority of machining centers. The machining center canned cycles are fixed canned cycles designed for programming drilling, boring, reaming, counterboring, countersinking, or tapping series of holes at different locations.

When using canned cycles for hole machining, the programmer needs to specify the cutting data for the first hole only. Then the control repeats cutting data for any number of holes. This saves a great amount of programming time and makes for programs that are simpler and easier to read.

There are several canned cycles used on machining centers, and although they are not standardized, most cycles are used on the majority of machining centers. Following is a general expression for the machining center canned cycle:

N__ G__ G98/G99 X__ Y__ Z__ R__ Q__ P__ F__ L__;

Address Description:

N	Block number
G	Type of canned cycle
G98	Return to the **initial point level** to which the tool retracts after the programmed depth on the Z axis is reached. This is used to move the tool freely over the clamps and bolts.
G99	Return to the **R point level** to which the tool retracts after the programmed depth on the Z axis is reached. This is used to allow some clearance above the part when the tool is moving between the holes.
X,Y	Hole center location
Z	Cutting depth
R	Point to which the tool retracts after the cutting depth on the Z axis is reached, used with a plus or minus sign
Q	Depth of each peck on the Z axis when programming a peck drilling cycle
P	Dwell
F	Feed rate
L	Number of repeats

All canned cycles on machining centers work on the same principle (Figure 8–2).

At the start of the canned cycle, the tool is brought in rapid to the first hole position specified by coordinates X1 and Y1. The tool continues in rapid until it reaches the R level. The tool then moves by the specified feed rate until it reaches the programmed hole depth on the Z axis. The tool retracts in rapid to the R point level or initial point level. The machining of the first hole is finished, and the tool moves to next hole location specified by the coordinates X2 and Y2.

The canned cycle is effective until canceled by the G80 code or replaced by another canned cycle. Any canned cycle can be programmed either in the absolute or incremental mode. Cutter radius compensation cannot be used; in fact, when machining holes, it is not necessary.

The majority of machining centers use the following canned cycles: G81, G82, G83, G84, G85, G86, G87, and G89. Each of them has a particular purpose, but they are programmed in a similar manner.

Drilling

Drilling is a metal-cutting operation in which the lips of a drill cut metal and produce chips. To ensure that the drill will not wander from a hole location when drilling, the programmer should first program a center drill operation. This is especially true when drilling holes on hard material.

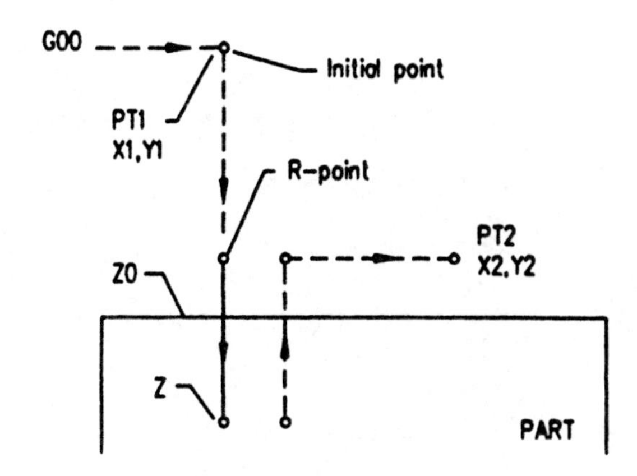

Figure 8–2 The canned cycle on machining centers.

On the blueprint, the designer usually specifies the **drilling depth** as a dimension from one reference surface to the end of the straight hole wall, such as 1.0 inch. If a drill with a point angle is used, the programmer must add the tool tip height to the blueprint size. For a 0.25 inch tool tip height, the programmer would enter 1.25 as the drilling depth. This value is usually programmed on the Z axis using a minus sign.

When the programmer must program a flat bottom hole, he or she should program the drilling depth as the blueprint size or less, using an end mill to finish.

If drilling is to be done on an angled surface, the programmer must program the end mill to cut a horizontal surface first. This enables the drill to penetrate the material without wandering.

When programming drilling on stainless steel, the programmer should use a slower RPM and a faster feed rate. The feed is also an important consideration when drilling small-diameter holes. Too much feed breaks the drill, but not enough feed makes the drill chatter, especially when drilling hard materials. Large drills tend to wander when penetrating hard material. This can be prevented by using a smaller size of drill first to make a path, then finishing the hole with the specified drill.

Deep hole drilling is programmed by reducing RPM about 20 percent and reducing feed rate about 10 percent. If the feed rate is reduced too much, the drill will chatter.

Drilling by the G81 Cycle

Programmers use the G81 cycle for drilling and center drilling. The tool starts from the R or initial point level, drills the hole until the programmed depth is reached, then retracts to the R or initial point level. The level to which the tool retracts depends on whether the programmer enters the G98 or G99 code as the tool return instruction. This cycle is illustrated in Figure 8–3.

Following is the instruction format for programming the G81 cycle:

G81 G99 (G98) X__ Y__ Z__ R__ F__;

Address Description:

G99 Return to the R point level
G98 Return to the initial point level
X,Y Hole location where the cycle starts
Z Drilling depth
R Point to which the tool retracts after the drilling depth is reached, used with a plus or minus sign
F Feed rate

Following is a programming example for the part illustrated in Figure 8–4. The material is mild steel.

Program:

G00 G90 G54 T1; (Coordinate system preset by the G54 function. Tool selection.)
M6; (Tool change.)
G0 X0 Y0 M08; (Rapid to the part center point where the part origin is set. Coolant on.)
G43 H1 Z1.0 S325 M03; (Approaching the part on the Z axis. Offset call. Spindle on.)
G81 G99 X4.242 Y4.242 R0.1 Z-1.5 F3.25; (Canned cycle specified to drill the first hole.)
X6.0 Y0; (Second hole location.)
X4.242 Y-4.242; (Third hole location.)
X0 Y-6.0; (Fourth hole location.)
X-4.242 Y-4.242; (Fifth hole location.)
X-6.0 Y0; (Sixth hole location.)
X-4.242 Y4.242; (Seventh hole location.)
X0 Y6.0; (Eighth hole location.)
G80; (Cycle cancel.)

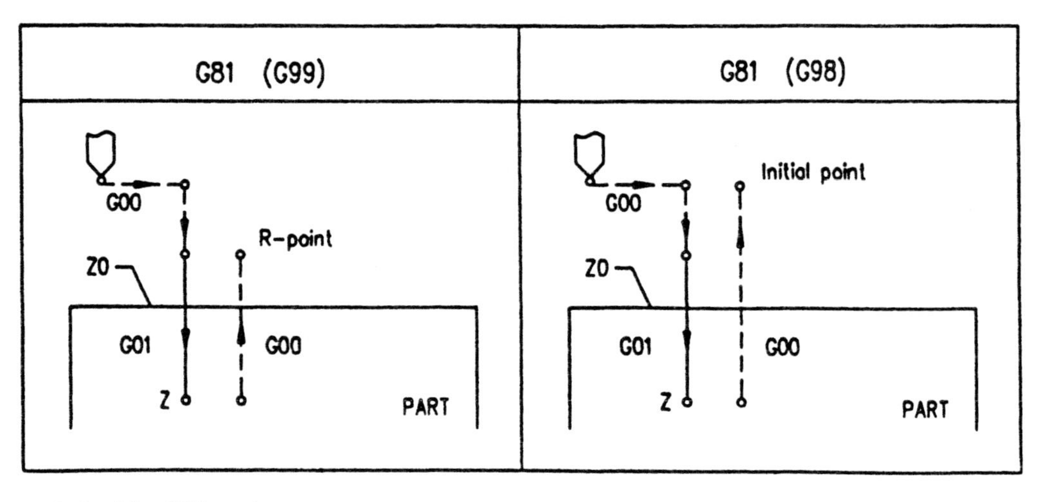

Figure 8–3 The G81 cycle.

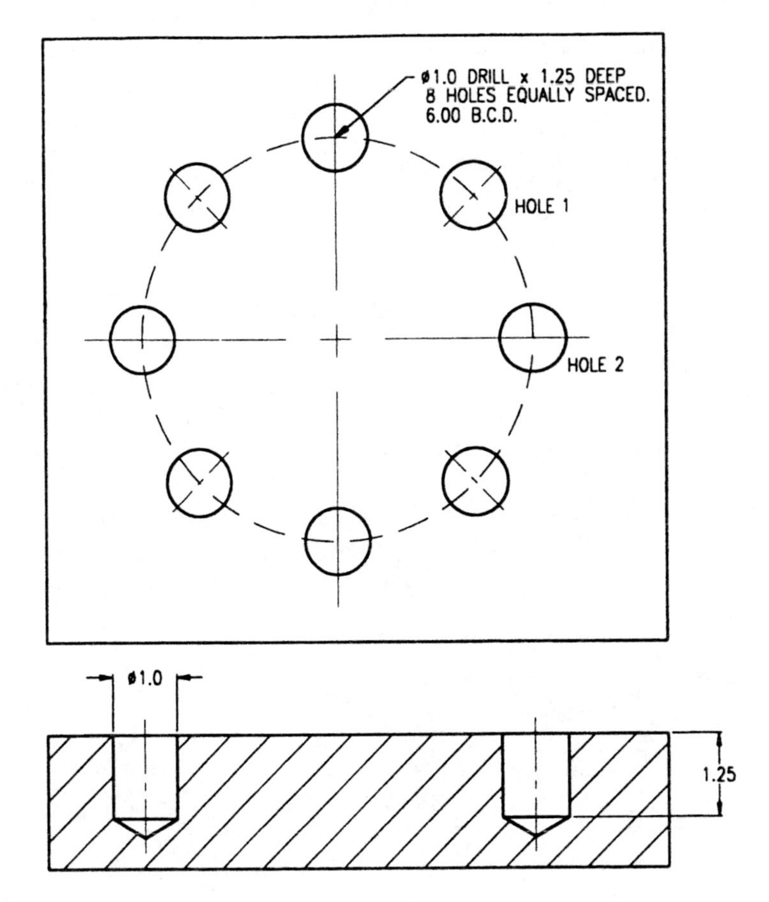

Figure 8–4 Drilling example.

```
G0 Z3.0 M9; (Retract on Z. Coolant stop.)
G28 Z0; (Return home on Z.)
M1; (Optional program stop.)
```

Notice that the drilling depth is programmed as 1.5 inch, not 1.25 inch as on the drawing. This is because the drill tip must be programmed deeper in order to cut a 1.25 inch straight wall as requested on the drawing.

Drilling with Dwell

The G82 cycle has the same function as the G81 cycle, but it dwells the motion when the drilling depth is reached. After the dwell, the tool retracts to the R or initial point level, depending on whether the programmer uses the G99 or G98 code as a tool return instruction (Figure 8–5).

The G82 cycle is used for drilling, counterboring, and countersinking. The dwell helps to get a better finish on the hole bottom and a more accurate Z depth dimension.

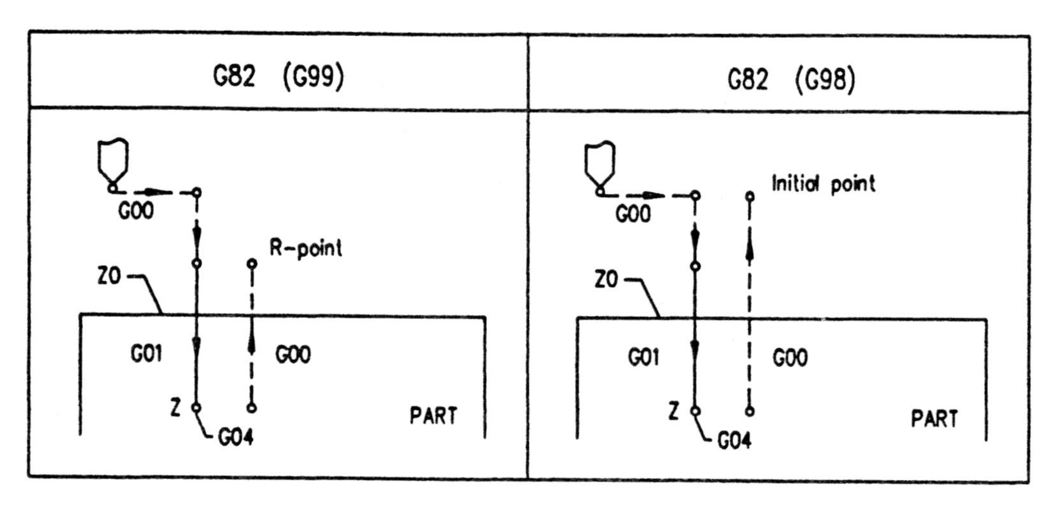

Figure 8–5 Drilling with dwell.

Counterboring is performed after a hole is drilled to open the beginning of the hole to a larger diameter. The purpose of counterboring is to clean a surface around the hole for the head of a bolt. The counterboring depth is shallow, usually 1⁄16 inch. Programmers can use a variety of tools for this operation, such as end mills, shell end mills, carbide-insert drills, and so on. The programmed speed and feed rate depend on the tool being used.

Countersinking operation adds the chamfer to a previously machined hole using a countersink tool with an 82-degree included angle. This operation should be programmed using about 60 percent of the drilling speed. To achieve a better finish when machining, the programmer may program the dwell, and the operator should use cutting oil.

Following is the instruction format for programming the G82 cycle:

G82 G99 (G98) X__ Y__ Z__ R__ P__ F__;

Address Description:

G99 Return to the R point level
G98 Return to the initial point level
X,Y Hole location where the cycle starts
Z Drilling depth
R Point to which the tool retracts after the drilling depth is reached, used with a plus or minus sign
P Dwell time in seconds or milliseconds
F Feed rate

Drilling with dwell is illustrated in the following series of instructions to drill the first three holes on the part shown in Figure 8–4.

G82 G99 X4.242 Y4.242 R0.1 P1 Z-1.5 F3.25; (Canned cycle specified to drill the first hole. Dwell time of one second is programmed by the address P.)

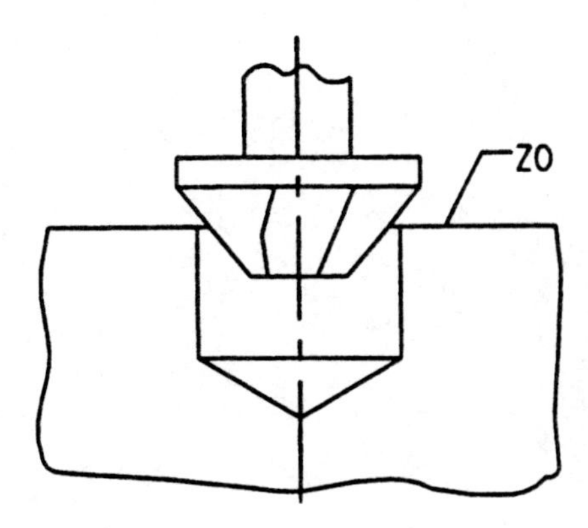

Figure 8–6 Setting up tapered tools.

X6.0 Y0; (Second hole location.)
X4.242 Y-4.242; (Third hole location.)
G80; (Cycle cancel.)

The operator can set up tapered tools such as the countersink in two ways. One way is to position the tool in the hole, as illustrated in Figure 8–6. The operator calculates the tool offset when the tool is in contact with the part surface by its cutting edge. The programmed depth is equal to the blueprint size of the countersink depth.

The other way to accomplish this is to set up the tool by touching off the part surface by its tip. The programmer must then take a taper angle into account. To calculate the tool travel on the Z axis to reach the hole with the tool cutting edge to make a taper or chamfer, the following formula may be used:

$$L = \frac{D - d}{2 \bullet \tan \alpha / 2}$$

L = Chamfer height
D = Chamfer diameter
d = Tool small end diameter
α = Included tool angle

To calculate the Z value when programming a 1.5-inch-diameter countersink using a 136-degree countersink tool with a 0.875 inch diameter on the small end:

$$L = \frac{D - d}{2 \bullet \tan \alpha / 2} = \frac{1.5 - 0.875}{2 \bullet \tan 136 / 2} = \frac{0.625}{4.95} = 0.126$$

To perform the same operations using the drill with a 118-degree angle, the programmed move in the Z direction may be calculated as:

$$L = \frac{D - d}{2 \bullet \tan \alpha / 2} = \frac{1.5 - 0}{2 \bullet \tan 118 / 2} = \frac{1.5}{3.328} = 0.450$$

Using this formula, the programmer can calculate drill tip height using a value of 0 for *d*.

Peck Drilling

The G83 is a peck drilling cycle similar to the peck drilling/grooving cycle on the lathe. This cycle differs from the G81 drilling cycle only by the Q code, which is used to program the amount of each peck in the Z direction. Pecking into material and rapid back continues until the specified drilling depth is reached. The retract amount depends on the parameter setting. The G83 cycle is illustrated in Figure 8–7.

The G83 peck drilling cycle is primarily used when drilling deep holes. A **deep hole** is a hole the depth of which is three times greater than the hole diameter. The retraction of the drill helps clear chips, which is important when drilling deep holes. Following is the instruction format when programming the G83 cycle:

G83 G99 (G98) X__ Y__ Z__ Q__ R__ F__;

Address Description:

G99	Return to the R point level
G98	Return to the initial point level
X,Y	First hole location
Z	Drilling depth
R	Point to which the tool retracts after the drilling depth is reached, used with a plus or minus sign
Q	Amount of each peck in the Z direction
F	Feed rate

Following is a programming example for the part illustrated in Figure 8–8. The material is steel 1020.

G00 G90 G54 T01; (Coordinate system preset by G54. Tool selection.)
M6; (Tool change.)

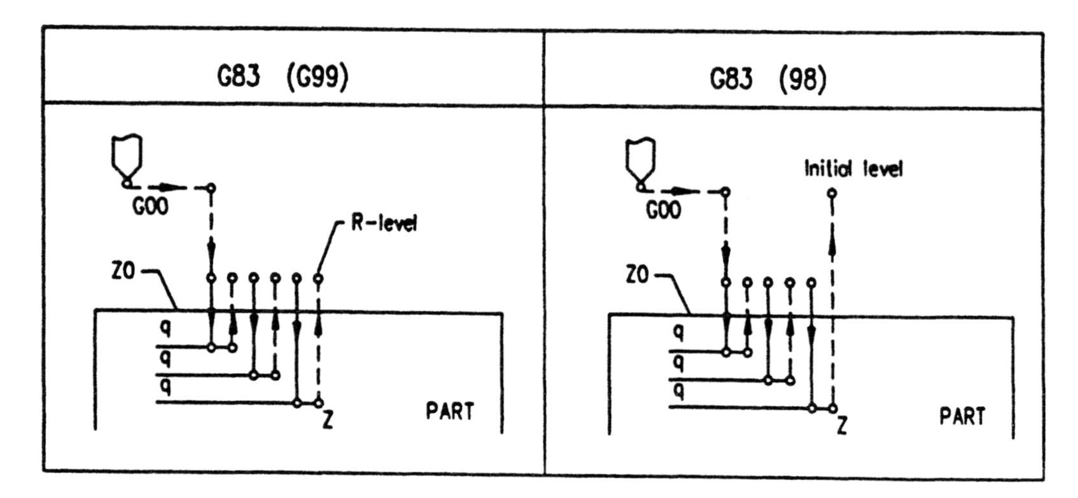

Figure 8–7 Peck drilling cycle.

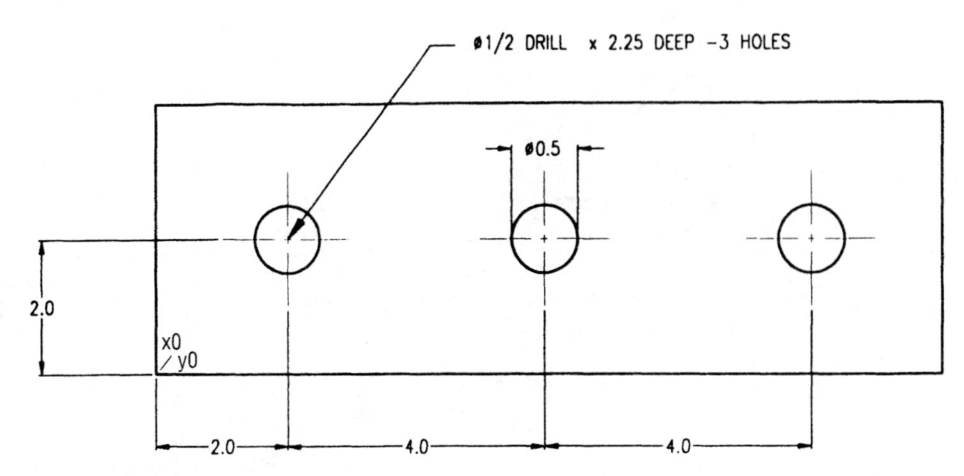

Figure 8–8 Peck drilling cycle example.

G0 X0 Y0; (Rapid to the part origin on the X and Y axes.)
G43 H1 Z1.0 S325 M03; (Approaching the part on the Z axis. Offset call. Spindle on.)
M08; (Coolant on.)
G83 G99 X2.0 Y2.0 R0.1 Z-2.4 Q0.15 F2.0; (Canned cycle specified to drill the first hole. The address Q specifies the amount of each peck in the Z direction.)
X6.0; (Second hole location.)
X10.0; (Third hole location.)
G80; (Canned cycle cancel.)
G0 Z3.0 M9; (Retract on Z. Coolant stop.)
G28 Z0; (Return home on Z.)
M1; (Optional program stop.)

The tool makes several pecks of 0.15 inch as specified by address Q. After each peck, the tool retracts to break the chips. Drilling of this hole continues until the programmed depth is reached. Then the drill moves to the position of another hole. The motion from the first to the second hole location, and from the second to the third hole location, is made only on the X axis. Thus, the programmer does not have to enter the Y address since the holes have the same Y value.

Tapping

Tapping is the operation of cutting the threads inside a previously drilled hole using a tap. The hole must be drilled smaller than the root diameter of the thread. There are several standards that recommend different combinations of tap and drill sizes.

The same cautions apply to programming and preparing a machine for tapping as for threading. During tapping, cutting oil should be used instead of coolant because it reduces heat better. Also, when tapping stainless steel, the operator should use cutting paste. Small holes, ¼ inch and less, should be programmed to start tapping on the machine but should be finished by hand tapping. This prevents breakage, especially when tapping on hard material. The machined thread start helps keep the tap straight when tapping by hand. Taps break for the following reasons:

- The drilled hole is too small.
- The hole is not drilled deep enough.
- The RPM is too high or too low.
- The tap and/or hole are not clean from chips.
- The tap and hole are not aligned.

On the blueprint, the **tapping depth** is specified for the length of full threads, such as 0.75. Since the tap tip cannot produce the full threads, the programmer must add the size of the tap tip height to the blueprint size. For a 0.25 inch tap tip height, the programmer enters 1.0 as the tapping depth. Sometimes the blueprint shows only tapping depth. Then the programmer and the operator have to calculate the drilling depth. For instance, if the tapping depth is 1.0 inch, the drilling depth may be calculated as follows:

1.0 (tapping depth, blueprint size) + 0.2 (tap tip height) = 1.2 inch
1.2 (tapping depth, calculated) + 0.25 (drill tip height) = 1.45 inch

Usually the programmer or the operator adds 0.01 inch to this value for clearance and enters 1.46 inch.

When programming a tapping operation, the programmer must use corresponding values for the feed rate and spindle speed in order for the hole to be tapped properly. When the machine is set to use the feed rate per spindle revolution, the pitch of the tap is programmed as the feed rate. For example, for a 0.125 inch pitch, the programmer would enter F0.125. Thus, the change in the spindle RPM amount would not affect the thread pitch.

When the machine is set to use the feed rate per time, the programmer can use the following formula to calculate the feed rate:

$$F = RPM \cdot P$$

The feed rate for a 0.125 inch pitch when using 130 RPM is calculated as:

$$F = 130 \cdot 0.125 = 16.25$$

Thus, the programmer would enter F16.25. Note that if the spindle RPM is changed, the feed rate also must be changed accordingly.

On a majority of machining centers, tapping is programmed by the G84 cycle. During this cycle, the tool performs several steps automatically (Figure 8–9).

Before the tapping operation starts, the tool moves in rapid to the hole position, then to the R point level. From the R point level the tool is moved by the programmed feed rate. When cutting a right-hand thread, the spindle turns in a clockwise direction. When it reaches the programmed depth, the tool stops and the spindle is reversed.

Then the tool is moved out from the hole by the same feed rate. When the tool reaches the R point level, the spindle is again turned clockwise for tapping the next hole. Then the tool moves to the new hole location.

Following is the instruction format for programming tapping on a machining center:

G84 G99 (G98) X__ Y__ Z__ R__ F__;

Address Description:

G99	Return to the R point level
G98	Return to the initial point level
X,Y	First hole location
Z	Drilling depth
R	Point to which the tool retracts after the tapping depth is reached, using a plus or minus sign
F	Thread pitch or lead

Following is a programming example for a tapping operation on the part illustrated in Figure 8–10. The material is aluminum bronze.

A subprogram can be used to specify the hole coordinates and call up the drilling and tapping cycle:

N100; (Drilling.)
N1 G00 G90 G54 X0 Y0; (Coordinate system preset. Rapid on X and Y to the part center point where the part origin is set.)
G43 H1 Z1.0 S375 M03; (Approaching the part on the Z axis. Offset call. Spindle on.)
G81 G99 R0.1 Z-1.5 F3.25; (Drilling cycle is not activated unless X or Y is specified.)

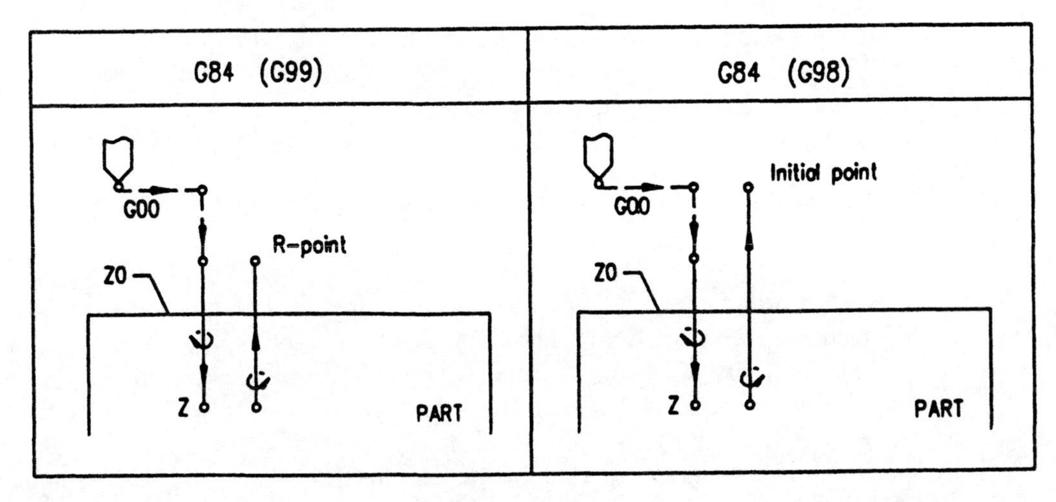

Figure 8–9 The tapping operation on machining centers.

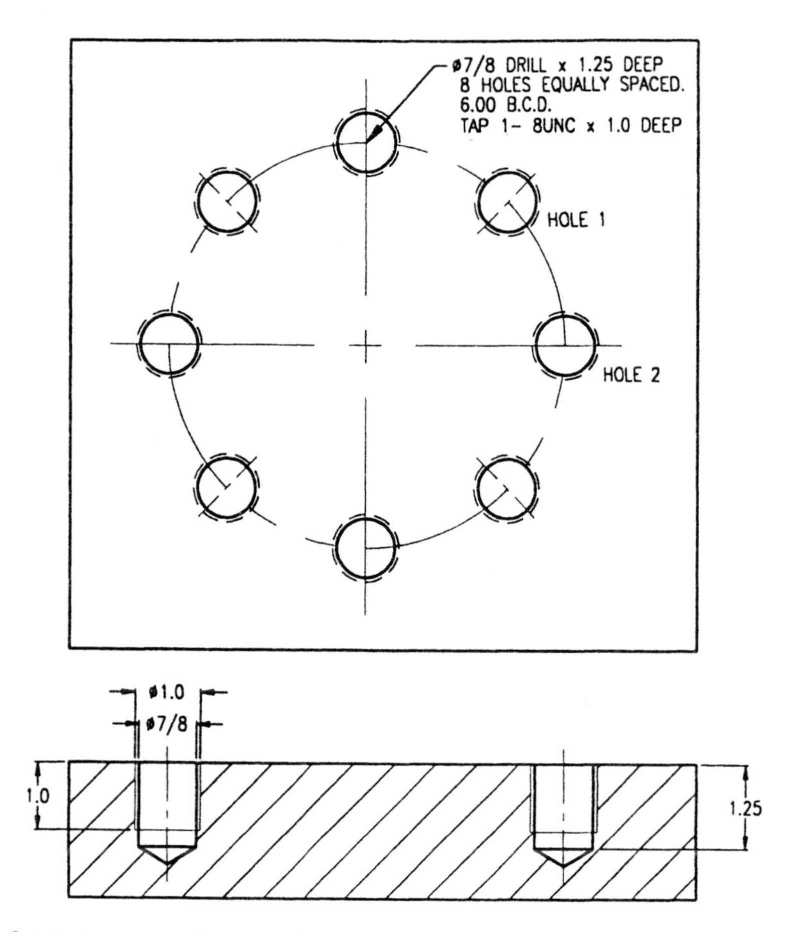

Figure 8–10 Programming a tapping operation.

M98 P30; (Calling the subprogram which specifies the hole coordinates.)
G00 Z3.0; (Cycle cancel; retract on Z.)
G28 Z0 (Return home.)
T2; (Calling T2 in waiting position.)
M6; (Tool change. T2 is loaded into the spindle.)
M1; (Optional program stop.)

N200; (Tapping.)
N2 G00 G90 G54 X0 Y0; (Coordinate system preset. Rapid on X and Y to the part center point where the part origin is set.)
G43 H2 Z1.0 S130 M03; (Approaching the part on the Z axis. Offset call. Spindle on.)
G84 G99 R0.1 Z-1.2 F16.25; (Tapping cycle.)
M98 P30; (Calling the subprogram which specifies the hole coordinates.)
G00 Z3.0; (Retract on Z.)
G28 Z0; (Return home.)

T1; (Calling T1 in waiting position.)
M06; (Tool change; T1 is loaded into the spindle, ready for program repetition.)
M30; (Program stop.)

(Subprogram)

O30; (Subprogram number.)
X4.242 Y4.242; (First hole location.)
X6.0 Y0; (Second hole location.)
X4.242 Y-4.242; (Third hole location.)
X0 Y-6.0; (Fourth hole location.)
X-4.242 Y-4.242; (Fifth hole location.)
X-6.0 Y0; (Sixth hole location.)
X-4.242 Y4.242; (Seventh hole location.)
X0 Y6.0; (Eighth hole location.)
M99; (Return to main program.)

Notice that in this program, the tool is not called at the beginning. It is assumed that the first tool was loaded into the spindle manually or by MDI, although the machine is equipped for a full random selection. The programmer can program the tool change in several ways. Following are some examples:

Example 1

G28 Z0; (Return the tool to the machine origin on Z axis.)
T3; (Tool selection.)
M6; (Tool change.)
M1; (Optional program stop.)

Tool selection and tool change instructions are programmed and executed separately. The operator has time to observe whether he or she has selected the proper tool.

Example 2

G28 Z0; (Return the tool to the machine origin on Z axis.)
T3 M6; (Tool selection by T3. Tool change by M6.)
M1; (Optional program stop.)

Tool selection and tool change instructions are programmed in the same block and executed at the same time. The operator has little no or time to see if the proper tool is selected.

Example 3

N1 G0 G90 G54 H1 T2; (Coordinate system preset for the first operation. Offset call for Tool 1. Tool 2 in waiting position.)
...; (Machining using Tool 1.)
G28 Z0; (Return Tool 1 to the machine origin on Z axis.)
M6; (Tool change; the machine takes T1 from the spindle and puts in T2 in waiting position selected in Line NI.)
M1; (Optional program stop.)

N2 G0 G90 G54 H2 T1; (Coordinate system preset for the second operation. Offset call for Tool 2. Tool 1 in waiting position.)
...; (Machining using Tool 2.)
G28 Z0; (Return Tool 2 to the machine origin on Z axis.)
M6; (Tool change; the machine takes T2 from the spindle and puts T1 in waiting position. The operator can continue machining using Tool 1 while machine is rotating the carousel looking for the pocket to place T2.)
M1; (Optional program stop.)

The machine selects the second tool while machining the part using the first tool. Note that in the beginning, the operator must load the first tool manually. Afterwards, the machine performs all tool changes automatically. Using this concept of programming the tool change, the programmer can cut the cycle time significantly, especially when the program calls for a greater number of tools.

The tool selection concept presented in Example 3 is a full random memory system available only on newer controls. In this system the tools do not necessarily return to the original tool pockets after use, meaning that the operator cannot easily locate the particular tool in the magazine. This is especially true when there are twenty or more tools loaded. If an operator wants to confirm the location of a particular tool, he or she enters the tool number using MDI and presses the CYCLE START button; the tool will rotate to waiting position. To make the tool change manually, the operator enters the following command using MDI:

G28 Z0; (Return the machine to its origin on Z.)
Txx M6; (Select the particular tool and make the tool change.)

Boring

Boring improves the surface finish and cuts the part to size. The tool follows its own path, fixing any deviation done by drilling. **Chatter** or **vibrations,** which reduce tool life and sometimes make it hard to obtain a good surface finish, may be a problem when boring. When attempting to eliminate vibrations, one should know what causes them. In general, vibrations are forced by external or internal factors.

Internal vibrations are caused by the instability of the cutting tool and workpiece as a system. Any tool deflection in the cutting process caused by a different part hardness or unevenness in heat treatment may produce chatter. Internal chatter is most common during boring operations.

External vibrations are caused by electric motors, compressors, and other machines working nearby, as well as the forces on the particular machine tool as a product of some machine parts, such as spindle bearings or gears, being worn out.

Some common troubleshooting methods for rough boring problems are:

- Check the insert and change it if necessary; secure the insert and boring bar.
- Increase the speed.
- Increase the depth of cut.
- Increase the feed rate for stainless steel.
- Check if the boring bar overhang can be shortened.
- Shim the boring bar to bring the tool tip above the centerline for the amount of the tool nose radius to get more clearance.

When a finishing boring bar chatters, the operator should check the finishing allowance (usually from 0.010 to 0.020 inch) or increase the speed. For deep boring with boring bars that must have a small cross-section, **devibration boring bars** should be used. They are made from a special material and double stress relieved, which reduces chatter. Solid carbide bars are also a big help.

There are several boring cycles—G85, G86, G87, G88, and G89—that work in a similar way: When the tool reaches the programmed depth on the Z axis, it retracts to the R or initial point level, and then the tool moves to another hole position.

The G85 cycle, similar to the G81 cycle, is used for rough boring, drilling with small drills, and reaming. Like boring, **reaming** improves the surface finish and cuts the part to size, but, unlike boring, the tool does not follow its own path: The reamer will repeat any deviation done by drill.

The programmer should choose a proper size drill prior to reaming. Too much material for reaming may result in reamer breakage and part damage. Holes under 1.000 inch in diameter are usually drilled 1⁄64 inch undersize prior to reaming. Holes over 1.000 inch in diameter are usually drilled 1⁄32 inch undersize prior to reaming. To program reaming, the programmer should use 60 to 75 percent of the drilling speed and feed rate. When pulling the tool out of the hole, he or she should not change the spindle rotation. Otherwise, the tool will leave marks inside the hole. Before reaming the hole, the operator should clean the reamer and the hole, using cutting oil to reduce heat while reaming.

During the G85 cycle, when the tool reaches the programmed depth, it returns to the start point without stopping or changing the spindle rotation, which is important for the reaming operation. On most machines, this cycle feeds out of the hole (Figure 8–11).

Following is the instruction format for programming the G85 boring cycle:

G85 G99 (G98) X__ Y__ Z__ R__ F__;

Address Description:

G99	Return to the R point level
G98	Return to the initial point level
X,Y	First hole location
Z	Boring or reaming depth
R	Point to which the tool retracts after the boring depth is reached, used with a plus or minus sign
F	Feed rate

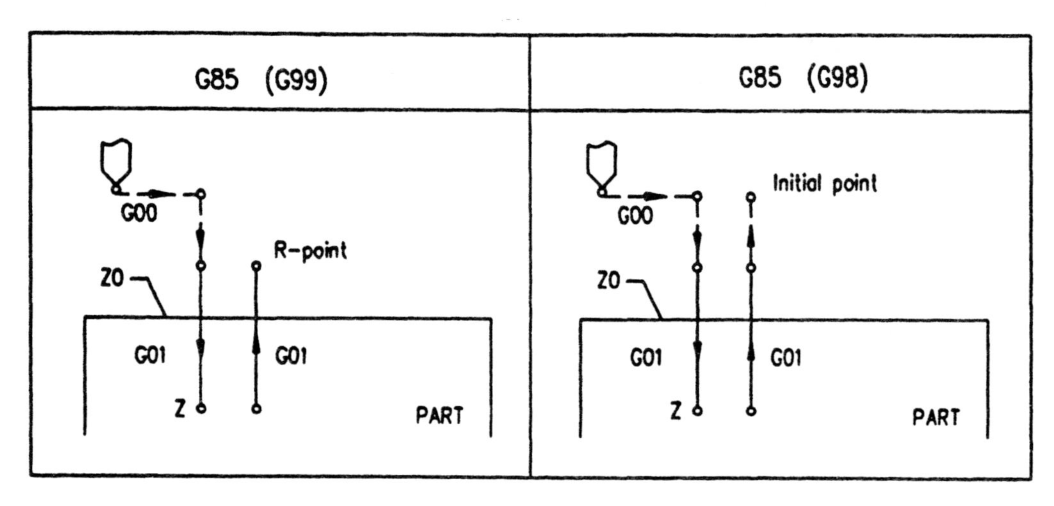

Figure 8–11 The G85 boring cycle on the machining center.

Following is a programming example for the boring operation on the part illustrated in Figure 8–12. The material is steel 4140.

G00 G90 G54 X0 Y0 TI; (Coordinate system preset by the G54 function. Rapid to the first part corner where the part origin is set.)
M6; (Tool selection.)
G43 H1 Z1.0 S125 M03; (Approaching the part on the Z axis. Offset call. Spindle on.)
M08; (Coolant on.)
G85 G99 X5.0 Y4.0 R0.1 Z-2.25 F1.0; (First bore location.)
X12.0; (Second bore location.)
G00 Z3.0 M09; (Cycle cancel; retract on Z. Coolant stop.)
G28 Z0; (Return home. Coolant stop.)
M1; (Optional program stop.)

Some special boring cycles are programmed in the same fashion as the G85 cycle. These are:

- G86 cycle for fine boring. This cycle is similar to the G81 and G85 cycles, except the spindle stops when the programmed depth is reached and returns in rapid to the cycle start point.
- G87 cycle is used for back boring or back facing. It may also be used to machine a bore undercut.
- G88 is the same as G87, except that when the programmed depth is reached, a dwell is performed for the amount of time specified by the P address. It is used for back boring with dwell.
- G89 cycle is the same as the G85 cycle, except that when the programmed depth is reached, a dwell is performed for the amount of time specified by the address P. It is used for ordinary boring with dwell.

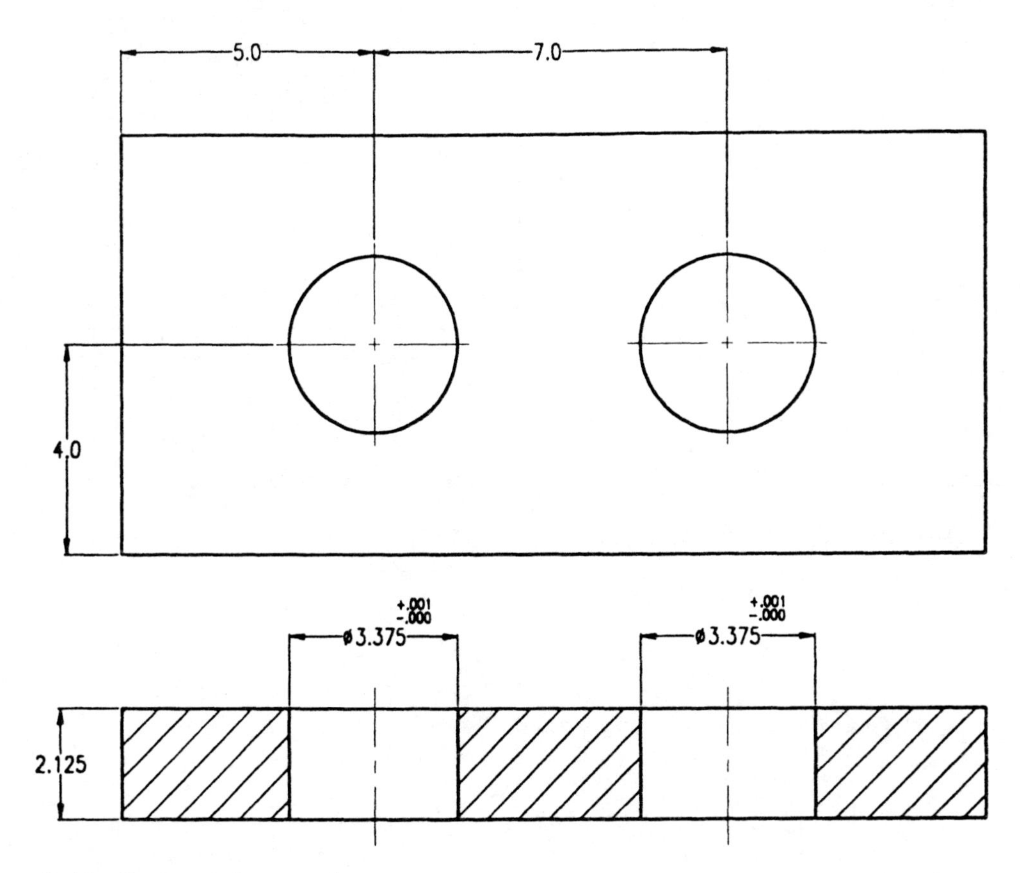

Figure 8–12 Boring cycle example.

Machining at Equal Intervals

Any machining center canned cycle can be used for machining holes at equal intervals. Once the cycle is specified for one hole location, it can be repeated for any number of holes if they are at equal intervals. The number of repeats is programmed using the L address. The maximum value for the number of repeats is usually a large number. Machining at equal intervals is illustrated in Figure 8–13.

Following is the instruction format for programming machining at equal intervals:

```
G81 G99 (G98) X__ Y__ R__ F__;
G91 X__ Y__ L__;
```

The canned cycle is programmed for the first hole only, either in the absolute or incremental mode. In the next block, the next hole location is specified by the X and Y addresses using the incremental mode and the G91 code. In the same block, the number of repeats is entered by address L. Following is a programming example for the part illustrated in Figure 8–14. The plate is ½ inch thick. The material is aluminum.

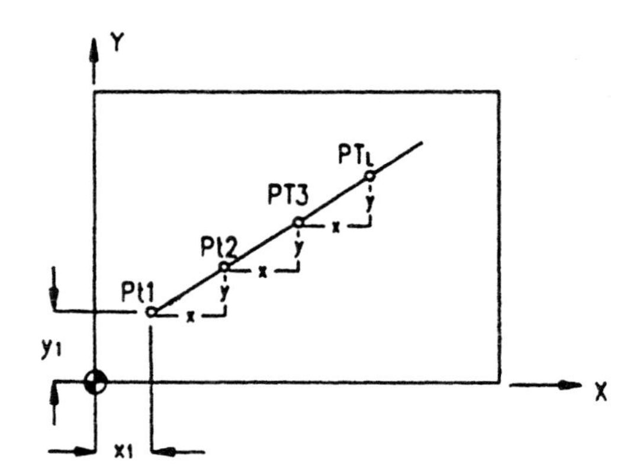

Figure 8–13 Machining at equal intervals.

G00 G90 G54 T01; (Coordinate system preset. Tool selection.)
M6; (Tool change.)
G0 X0 Y0; (Rapid to position on the X and Y axes.)
G43 Z1.0 H1 S850 M03; (Approaching the part on the Z axis. Spindle start.)
M8; (Coolant on.)
G81 G99 X1.0 Y1.0 Z-0.65 R0.1 F2.7; (Canned cycle specified for the first hole location.)
G91 X0.707 Y0.707 L6; (Next hole location in incremental programming.)
G00 G80 G90 Z1.0; (Canned cycle cancel. Return to absolute programming.)
X4.535 Y1.0; (Positioning to a new location.)
G81 G99 Z-0.65 R0.1 F2.5; (Drill the first hole at a new location.)
G91 Y0.5 L4; (Incremental programming.)
G80 G90; (Canned cycle cancel. Return to absolute programming.)
G0 Z2.0 M09; (Retract on Z. Coolant stop.)

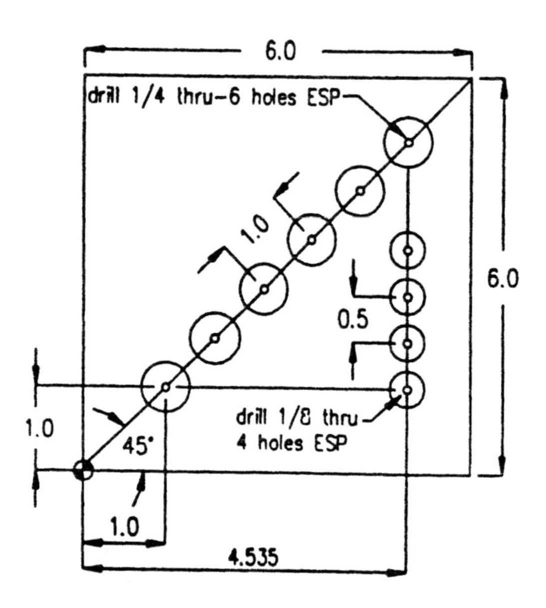

Figure 8–14 Machining at equal intervals example.

```
G28 Z0; (Return home.)
M01; (Optional program stop.)
```

For the first hole pattern, the displacement on the X and Y axes is the same because of a 45-degree angle. It is calculated using the sine or cosine function from the right triangle, the hypotenuse of which is 1.0 inch. For the second hole pattern, the displacement is programmed in the Y direction only because the holes are in a straight line perpendicular to the X axis.

Milling

Milling is a process of removing metal using multipoint tools. It is both a roughing and a finishing process that produces precise contours and shapes. Milling applications fall into two general categories:

1. Side milling: straight or contour milling, roughing or finishing
2. Milling multiple surfaces: three-dimensional profile milling, cutting a keyway, thread milling

The milling operation produces strong forces that may cause vibration and deflection of either the part being machined, the work-holding device, the cutting tool, or the machine itself. To prevent or minimize vibration, the part must be securely held in a work-holding device capable of resisting the forces involved when machining. Cutter runout, spindle runout, the alignment of the cutter and the spindle, and the rigidity of the workpiece, fixture, tool, and machine must all be considered when attempting to predict the accuracy of milling operations. Therefore, the setup must be rigid (Figure 8–15).

On any type of operation, when the part must be machined in close tolerance, the programmer should design the program for the roughing and finishing tool. This is especially true for milling operations. When the programmer uses the same tool for both roughing and finishing, the tool is already partially dull as the operator starts the finishing operation. Consequently, the finish will not be as good as when the programmer uses roughing and finishing tools.

Some machining centers have fixed canned cycles for face milling and pocket milling. Since they are not suitable when milling complex part shapes, the programmers use automatic tool radius compensation or part programming software to program milling operations.

When programming by automatic tool radius compensation, the programmer programs the tool to the points of intersection. Literally speaking, the programmer programs the points on the part, not the tool reference point. Then the tool automatically shifts away from the part. The shift amount depends on the value stored in the geometry offset using MDI. Geometry offsets are used for the following purposes:

Figure 8–15 Typical milling operation. To prevent or minimize vibration, proper tooling is one of several factors to be considered for successful and safe operation. (Courtesy Valenite, Inc.)

1. Using a different tool diameter than the one originally programmed
2. Varying the part size while machining
3. Making a series of cuts when roughing and finishing using the same programmed data
4. Adjusting for tool wear

Automatic tool radius compensation makes it possible to use a different tool diameter instead of the one originally programmed. For example, say a program calls for a 1.000-inch-diameter cutter, a size that is not available. The closest available size is 1.0625 inch diameter. The operator would enter a 0.5312 radius value instead of 0.5000 into the particular geometry offset. The control would shift the tool away from the part for this amount and the part would be machined as if the original cutter size was used. No change in the program would be needed since the control would recalculate the tool path according to the value in the geometry offset.

When using automatic tool radius compensation, the operator can easily vary the part size while machining. For instance, if a part is machined oversize [Figure 8–16, position (a)], he or she should decrease the geometry offset value by the appropriate amount. If the part is 0.010 inch oversize, when using a 1-inch-diameter cutter, the operator should change the R value entered from 0.5000 to 0.4990, or he or she should enter –0.010 into geometry offset. Either way will shift the tool

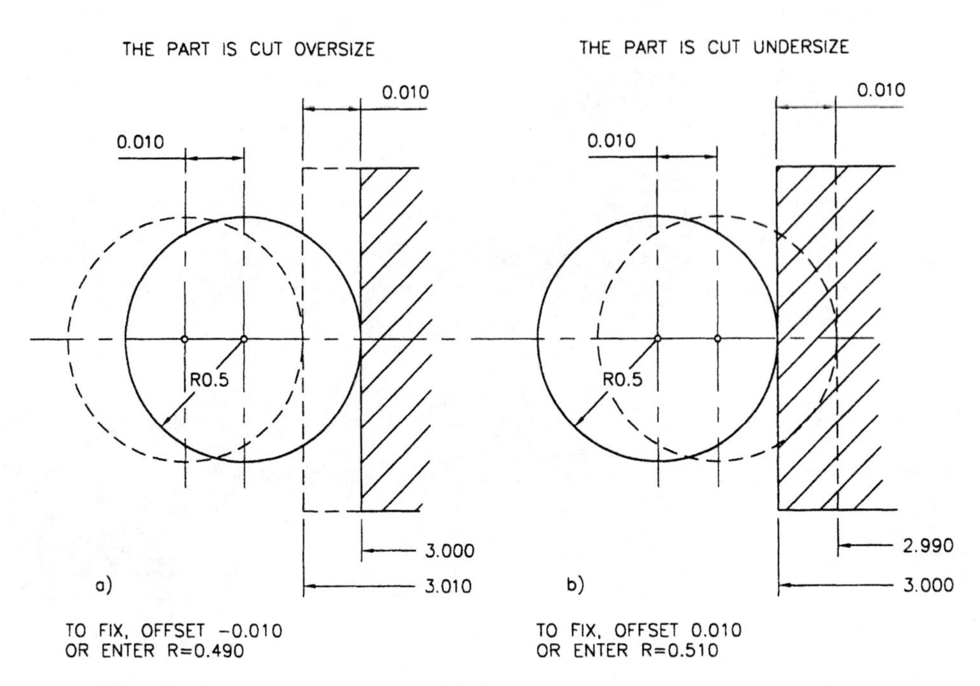

Figure 8–16 Sizing the part when machining.

0.010 inch closer to the part, producing the desired size. The operator should do the opposite if the part is machined undersize [position (b)]. Then he or she should increase the R value (making the control think a larger cutter is used even if it is not) or he or she should enter 0.010 into geometry offset. Either way will shift the tool 0.010 inch away from the part, producing the desired size.

Note that on some machines there are two types of geometry offset, wear and geometry. The wear offset is used for fine adjustment for tool wear, while the geometry is used to enter the size of the tool radius. The operator can use either one to affect the part size, but not both at the same time.

By using different geometry offsets, the programmer can use the same programmed data to make a series of cuts when roughing and finishing. For example, say for the roughing operation he or she assigns a 1.010 inch tool radius to offset 01, and for the finishing operation a 1.000 inch tool radius to offset 11. When using offset 01, the tool will leave 0.010 inch for finishing. The part will be cut to a finished size when using offset 11. Thus, the programmed data is the same: Only the offset value is different.

In order to enable the control to use the tool radius compensation, the programmer and the operator should do two things:

1. The programmer should call the particular geometry offset in the program when initializing the compensation.
2. The operator should enter the size of the tool radius into the particular geometry offset using MDI.

On the machining center, the address D is used to call the particular geometry offset in the program. (Note that some machines use R for this purpose.) This allows the programmer to label the geometry offset number differently than the tool length offset number. For example, he or she may assign H2 to the tool length offset for one particular tool and D31 to the geometry offset for the same tool. This important feature of the machining center allows the programmer to call and cancel one type of offset without affecting the other. To make the machine ready to use tool radius compensation, the operator enters the size of the tool radius into the tool offset register using address D. The programmer must call or cancel tool radius compensation for one particular tool as many times as needed, but the operator must enter the size of the tool radius for that tool only once.

Once given the appropriate compensation code, the tool changes its position to the compensated position on the next programmed move. It continues cutting in the compensation mode until the compensation is canceled. On the next programmed move, the tool position is changed from the compensated to the uncompensated position. Before returning the tool to the home position, the compensation must be canceled. If not done, the control remembers the compensation amount and uses it later in the program. This may cause the following problems:

1. Compensation may be used for a tool that does not need it, such as a drill.
2. The compensation amount may be built on the compensation amount for the next tool.
3. Some machines may display an alarm.

The compensation is initialized or canceled in the program when the control reads the following codes (which have the same purpose as on the lathe):

G41 Compensation left
G42 Compensation right
G40 Compensation cancel

When executing the program, the control reads several lines at once in order to position the tool for the next move. This is important when automatic tool radius compensation is used, since the tool radius must always be set perpendicular to both surfaces, present and upcoming. Reading the program in several blocks helps the control not only to execute the instructions, but also to find the erroneous ones in order to prevent their execution and to generate the alarm.

When canceling compensation, the programmer should make sure that the tool has room to retract; otherwise, overcutting may occur. If there is no room for the tool to retract, the programmer should move the tool to a safe distance before canceling the compensation. Afterwards, he or she can cancel the compensation at any convenient point, usually while the tool is returning to the home position.

When programming using automatic tool radius compensation, always keep in mind that once initialized, compensation stays in effect until canceled. This means that the next tool will use the compensation even if it is not programmed. If this tool also calls the compensation, it will build on the previous one. If a tool is programmed without compensation (drill or tap, for instance), it may behave in a strange way. Thus, the programmer should always be sure to cancel the compensation. Otherwise, it may be as negative as if the tool offset is not canceled. To make sure that the compensation is canceled, some programmers enter the G40 code in a safety line at the beginning of the program. For example,

N1 G00 G40 G80;

In this line, G40 cancels the compensation if programmed before, while G80 cancels any canned cycle previously programmed.

Following is a programming example for the milling step on the part shown in Figure 8–17. The material is $3\frac{1}{2} \times 2\frac{1}{2} \times 1$ inch aluminum plate.

The order of machining operations:

Tool 1, Offset H01; 4.0-inch-diameter face mill, to mill the top to size
Tool 2, Offset H02 and D31; 1-inch-diameter roughing end mill
Tool 3, Offset H03 and D32; 1-inch-diameter finishing end mill

On the Z axis, the part origin is set at the part finish face, so the part must be milled off about 0.125 inch. The operator should use the tool length offsets (H offsets) to adjust on the Z axis, and geometry offsets (D offsets) to adjust on the X and Y axes. For the end mill, two geometry offsets will be used; D31 for roughing and D32 for finishing.

After the setup of the tools and prior to starting to machine the first part, experienced operators increase the offset values from 0.010 inch to 0.030 inch. This moves

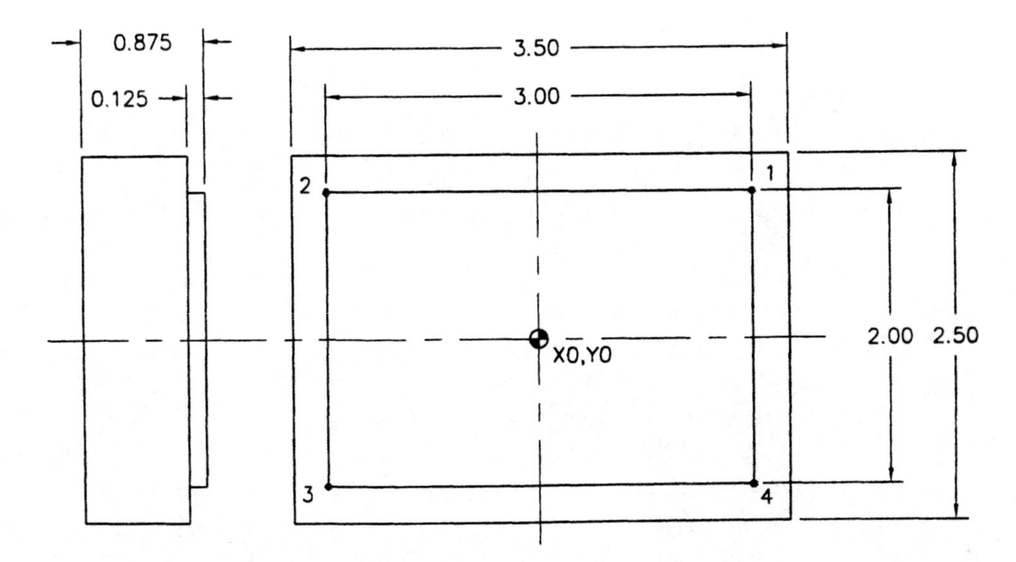

Figure 8–17 Milling example.

the tools away from the part, so the tools cut less. After measuring the part, if needed, the operator adjusts the offsets and re-runs the program to cut the part to proper size.

Program:

```
N1 (Milling the top of the part using 4.0-inch face mill.)
G0 G90 G54 S750 T01 M03; (Absolute programming. Coordinate system
    preset. Tool 1 in ready position. Spindle on at 750 RPM.)
M06; (Loading Tool 1.)
G0 X-4.0 Y0; (Rapid to start position on the left-side of the part.)
G43 H01 Z0; (Lowering the tool.)
G01 X4.0 F12.0; (Milling.)
G0 Z2.0; (Retract on Z.)
G28 Z0; (Return to the machine origin on Z.)
M01; (Optional program stop.)

N2 (Roughing.)
G0 G90 G54 S950 T02 M03; (Absolute programming. Coordinate system
    preset. Tool 2 in ready position. Spindle on at 950 RPM.)
M06; (Loading Tool 2.)
G0 X2.35 Y1.85; (Rapid to start position at the right-top corner, calculated as
    3.5 / 2 + 0.6 = 2.35 for the X axis and 2.5 / 2 + 0.6 for the Y axis.)
G43 H02 Z-0.120; (Lowering the tool leaving a 0.005 inch for finishing.)
G01 G42 D31 Y1.0 F12.0; (Compensation right. Calling geometry offset D31.
    Positioning the tool on Y axis.)
X-1.5; (Cutting up to point 2.)
Y-1.0; (Point 3.)
X1.5; (Point 4.)
Y1.25; (To pass point 1.)
G0 Z2.0; (Retract on Z.)
G40; (Compensation cancel.)
G28 Z0; (Return to the machine origin on Z.)
M01; (Optional program stop.)

N3 (Finishing.)
G0 G90 G54 S1100 T03 M03; (Absolute programming. Coordinate system
    preset. Tool 3 in ready position. Spindle on at 1100 RPM.)
M06; (Loading Tool 3.)
G0 X2.35 Y1.85; (Rapid to start position.)
G43 H03 Z-0.125; (Lowering the tool.)
G01 G42 D32 Y1.0 F12.0; (Compensation right. Calling geometry offset D32.
    Positioning the tool on the Y axis.)
X-1.5; (Cutting up to point 2.)
Y-1.0; (Point 3.)
X1.5; (Point 4.)
Y1.25; (To pass point 1.)
```

```
G0 Z2.0; (Retract on Z.)
G40; (Compensation cancel.)
G28 Z0; (Return to the machine origin on Z.)
M01; (Optional program stop.)
```

The programmer can make this program shorter by using a subprogram. He or she can also speed it up by calling the tool for the next operation while the present tool is cutting. Then, when running the program for the first time, the operator must load the first tool manually. For example:

```
N1 (Milling the top.)
G0 G90 G54 S750 T02 M03; (While selecting Tool 2, the machine proceeds for
    machining using Tool 1. These two operations are simultaneous.)
G0 X-4.0 Y0; (Rapid to start position.)
G43 H01 Z0; (Calling offset H01 and lowering the tool.)
G01 X4.0 F12.0; (Milling.)
G0 Z2.0; (Retract on Z.)
G28 Z0; (Return to the machine origin on Z.)
M6; (Tool change. The machine is unloading Tool 1 and loading Tool 2.)
M01; (Optional program stop.)

N2 (Roughing.)
G0 G90 G54 S950 T03 M03; (While selecting Tool 3, the machine can proceed
    for machining using Tool 2. The machine does not have to wait for comple-
    tion of tool selection operation.)
G0 X2.35 Y1.85; (Rapid to start position.)
G43 H02 Z-0.120; (Lowering the tool.)
G01 G42 D31 Y1.0 F12.0; (Compensation right. Calling geometry offset D31.
    Positioning on the Y axis.)
M98 P150; (Subprogram call.)
G28 Z0; (Return to he machine origin on Z.)
M6; (Tool change. The machine is unloading Tool 2 and loading Tool 3.)
M01; (Optional program stop.)

N3 (Finishing.)
G0 G90 G54 S1100 T01 M03; (While selecting Tool 1, the machine can proceed
    for machining using Tool 3. These two operations are simultaneous.)
G0 X2.35 Y1.85; (Rapid to start position.)
G43 H03 Z-0.125; (Lowering the tool.)
G01 G42 D32 Y1.0 F12.0; (Compensation right. Calling geometry offset D32.
    Positioning on the Y axis.)
M98 P150; (Subprogram call.)
G28 Z0; (Return to the machine origin on Z.)
M6; (Tool change. The machine is unloading Tool 3 and loading Tool 1. The
    machine can continue machining using Tool 1, while the machine rotates
    the carousel looking for the pocket to place T3.)
M01; (Optional program stop.)
```

```
O150; (Subprogram number.)
X-1.5; (Point 2.)
Y-1.0; (Point 3.)
X1.5; (Point 4.)
Y1.25; (To pass point 1.)
G0 Z2.0; (Retract on Z.)
G40; (Compensation cancel.)
```

To be able to use the same subprogram and different tools and tool offsets, the programmer should enter the offset calls in a main program. Then, he or she can use a subprogram to repeat the coordinates on a tool path.

Helical Milling

Helical milling is accomplished by **helical interpolation,** another important feature in CNC programming. It allows circular interpolation on two axes simultaneously, while providing linear cutting on a third axis. Helical interpolation is most commonly used in milling large-diameter threads, grease grooves, helical pockets, and other helical forms.

All three axes move at the same time to produce the **helix** or **spiral.** When cutting the helix, the tool moves simultaneously on the X and Y axes to produce a circular interpolation, while at the same time it is moving in linear motion on the Z axis to achieve a correct pitch.

Helical interpolation is usually an optional feature purchased at additional cost. On a majority of machining centers, helical interpolation is initialized by the G14 code for clockwise or G15 code for counterclockwise interpolation. Then, the X and Y coordinates specify the arc end point, while the Z coordinate is used to program the length of thread. The I, J, and K addresses are used to describe the arc center displacement. The pitch of the helix is specified by the address F, which is the tool feed rate on the Z axis. The number of complete threads is programmed by the address L.

The following example illustrates how to program helical interpolation to cut a 60-degree external thread, 10 threads per inch, 0.065 inch deep on a 5.0-inch-diameter shaft (Figure 8–18). The thread is 1.5 inch long and the material is aluminum. A standard 3.0-inch-diameter cutter will be used.

Thread Data:

Root diameter: $D = 5.0 - (2 \bullet S) = 5.0 - (2 \bullet 0.065) = 4.870$
Number of full arcs: $L = 1.5$ (length of thread) $/ P = 1.5 / 0.1 = 15$

The programmer should add the length of at least one pitch for both acceleration and deceleration of the tool. Thus, 17 full arcs should be programmed. Assuming that the part origin is set at the center of the part, the value for the root diameter must be divided by 2 to get the tool start position according to the part origin. Thus,

$$4.870 / 2 = 2.435$$

Then the following coordinates are used to program the tool start point before a helical interpolation is initialized:

X4.1 (calculated as: 5.0/2 + 3.0/2 + 0.1 clearance.)
Y0 (The coordinate of the part origin on the Y axis.)
Z0.1 (Tool start point on the Z axis; one pitch value.)

Following is a complete program to machine a 1.5-inch-long thread using helical interpolation:

```
G00 G90 G54 T01; (Coordinate system preset. Tool selection.)
M06; (Tool change; loading Tool 1.)
G0 X4.1 Y0 S125 M03; (Approaching the part in rapid on the X and Y axes.
    Spindle start at 125 RPM.)
G43 H01 Z0.125; (Offset call. Approaching the part on the Z axis.)
G01 G41 D30 X2.435 F17.0 M8; (Feeding on the X axis to reach the coordinates
    of the root diameter. Compensation left. Coolant on.)
G14 X2.435 Z-1.6 I-2.435 F12.5 L17; (Helical interpolation.)
G01 X4.1 F17.0.0 M09; (Leaving the part by a faster feed.)
G0 Z1.0 M9; (Retract on Z. Coolant stop.)
G40; (Compensation cancel.)
G28 Z0; (Return home on Z.)
M1; (Optional program stop.)
```

If the programmer wants a finishing pass, he or she can increase the tool start point X coordinate in roughing by the amount of the finishing pass. If the programmer does not want to use tool radius compensation, he or she can add a tool

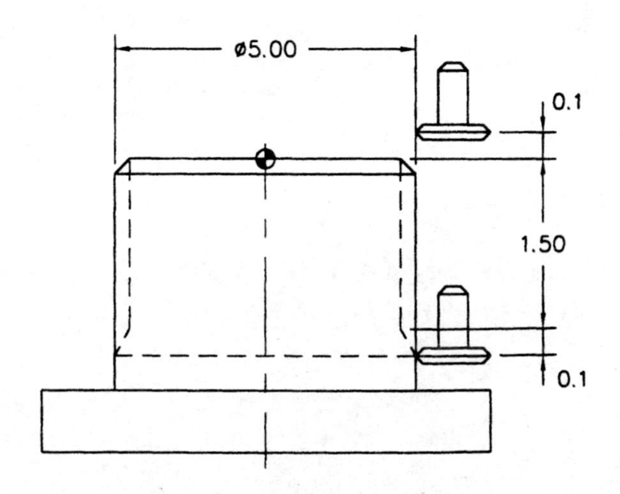

Figure 8–18 Helical interpolation example.

radius value to the X coordinate of the thread root diameter. For example, 2.435 + 1.5 (tool radius) = 3.935. Note that the I value does not have to be changed.

On machines without the helical interpolation feature, the programmer can program a helix if the machine has the **fourth axis,** known as the A or B axis. Then, circular interpolation is accomplished on the A or B axis, while linear tool motion is provided on the X axis. In this case, on a vertical machining center the part is placed horizontally. The following example is the program for the part shown in Figure 8–19. The material is mild steel.

The tool start point will be 0.375 inch from the face on the right-hand side of the part, where the part origin is set. The tool should stop cutting after it is past the part for the amount of tool radius plus tool clearance. Therefore, total tool travel is calculated as:

0.375 (tool start point) + 2.25 (length of the part) + 0.375 (tool exit point) = 3.00 inch

The pitch of helix is 0.75 inch. This is the amount of motion on the X axis after the part makes a full turn (360 degrees). To find the number of turns needed to make a helix on the length of the shaft, tool travel must be divided by the pitch of helix. Thus,

$$3.0 / 0.75 = 4$$

As calculated, four full turns of the work are needed to cut a helix on the length of 2.25 inches (adding tool clearance on the start and the end points). The

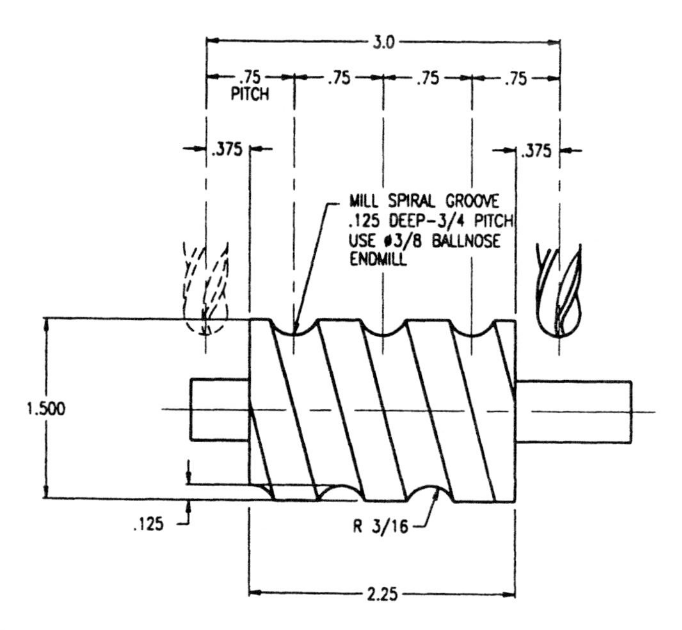

Figure 8–19 Example of helical milling on a machine without the helical interpolation feature.

programmer must convert the number of turns into degrees as required when programming the tool motion on either the A or B axis, depending on a particular machine/control combination, as follows:

$$360 \cdot 4 = 1440 \text{ degrees}$$

Following is a complete program to machine a helix for the part shown in Figure 8–19:

```
G00 G90 G54 T01 A0; (Coordinate system preset by the G54 function. Tool
    selection. Positioning A axis to zero-degree position.)
M6; (Loading Tool 1 into the spindle.)
G0 X0.375 Y0; (Approaching the part in rapid on the X and Y axes.)
G43 H1 Z1.0 S850 M03; (Approaching the part on the Z axis. Offset call.
    Spindle on.)
Z-0.125 M8; (Lowering the tool for the depth of cut on the Z axis.)
G1 X-2.625 A1440 F120.0 M08; (Cutting the helix. Coolant on.)
G00 Z2.0 M09; (Retract on Z. Coolant stop.)
G28 Z0; (Return home.)
M1; (Optional program stop.)
```

For the Y axis, the tool rests on the part centerline, while simultaneous linear motion on the X axis and rotation of the A axis cut the helix. The helix is cut using the feed rate of 120 degrees/minute as specified by address A. The part will make a complete turn in about three minutes. If linear motion on the X axis is not programmed, only a full arc (not a helix) will be cut. When a great number of passes are needed to finish a helix, a subprogram may be used.

Summary

Machining center canned cycles are fixed cycles. Once a canned cycle is specified, the cutting data programmed for the first hole can be repeated for any number of holes. This saves a great amount of programming time, and makes programs simpler and easier to read.

Canned cycles on machining centers are not standardized, but most of them, such as G81, G82, G83, G84, G85, G86, G87, G88, and G89, are used on the majority of machines. Programmers use these canned cycles for the following purposes: G81 for drilling and center drilling; G82 for drilling with dwell, counterboring, and countersinking; G83 for for peck drilling deep holes; G84 for tapping; and G85 for boring and reaming operations. The G85, G86, G87, G88, and G89 cycles are used for programming special cases of boring operations. Any machining center canned cycle can be used for machining holes at equal intervals.

Programmers normally use part programming software or automatic tool radius compensation for programming milling operations. This is especially true when they need to design programs for milling complicated part shapes such as molds and cams. In automatic tool radius compensation, the tool is programmed to the points of intersection. Then the tool is automatically shifted away from the part. The shift amount depends on the value stored in the geometry offset using MDI. The geometry offsets are used for the following purposes: using a different tool diameter than the one originally programmed; varying the part size while machining; making a series of cuts when roughing and finishing using the same programmed data; and adjusting for tool wear.

In order to enable the control to use the tool radius compensation, the programmer should call the particular geometry offset in the program when initializing the compensation, and the operator should enter the size of the tool radius into the particular geometry offset using MDI. The programmer must call or cancel tool radius compensation for one particular tool as many times as needed, but the operator must enter the size of the tool radius for that tool only once.

On the machining center, the address D is used to call the particular geometry offset in the program. This allows the programmer to label the geometry offset number differently than the tool length offset; he or she uses the H address for the tool length offset and the D address for the geometry offset number for the same tool. This is an important feature of the machining center, which allows a programmer to call and cancel one type of offset without affecting the other.

Once initialized, compensation stays in effect until canceled. This means that the next tool will use the compensation even if it is not programmed. If this tool also calls the compensation, it will build on the previous one. The programmer should always be sure to cancel the compensation. Otherwise, it may be as negative as if the tool offset is not canceled.

When the part must be machined in close tolerance, the programmer should design the program for the roughing and finishing tool using different geometry offsets. Then the operator can size the part to any tolerances requested by the blueprint. To accomplish this, he or she can simply change geometry offset value of the size of the tool radius.

Helical milling is accomplished by helical interpolation, which is another important feature in CNC programming. It allows circular interpolation on the two axes simultaneously, while providing linear cutting on a third axis, producing a helix or spiral. Helical interpolation is most commonly used in milling large-diameter threads, grease grooves, helical pockets, and other helical forms.

On a majority of machining centers, helical interpolation is initialized by the G14 code for clockwise or the G15 code for counterclockwise interpolation. On machines without the helical interpolation feature, the programmer can program a helix using the G01 function if the machine has a fourth axis, known as the A or B axis. Then, circular interpolation is accomplished on the A or B axis, while linear tool motion is provided on the X axis.

When programming helical interpolation, the programmer should take into account the acceleration and deceleration of the tool; the tool should be switched into the helical mode before touching the part. This is opposite when the tool is leaving the part; it should still be in helical mode for some time after the tool has left the part.

Key Terms

chatter
counterboring
countersinking
deep hole drilling
devibration boring bars
drilling depth
external vibrations
fourth axis
helical interpolation
helix
initial point level
internal vibrations
milling
reaming
R point level
spiral
tapping depth
vibration

Self-Test

The answers are in Appendix E.

1. Most of time, the _______________ is programmed by adding the drill tip height to the drilling depth specified on the blueprint.
2. When programming the _______________, the tap tip height should be added to the drilling depth.
3. _______________ is programmed by reducing the RPM about 20 percent and reducing feed rate about 10 percent.
4. To program _______________, 60 to 75 percent of the drilling speed and feed rate should be used.
5. _______________ operation should be programmed using about 60 percent of the speed for drilling.
6. Returning to the _______________ is programmed by the G98 code.
7. Returning to the _______________ is programmed by the G99 code.
8. _______________ reduces tool life and sometimes makes it hard to obtain a good surface finish.
9. _______________ are caused by the instability of the cutting tool and workpiece as a system.
10. For deep boring with boring bars that must have a small cross-section, _______________ should be used.
11. _______________ is a process of metal removal using multipoint tools.
12. To prevent or minimize _______________, the part must be securely held in a work-holding device.

13. _______________ allows circular interpolation on two axes simultaneously while providing linear cutting on a third axis.
14. When cutting a _______________, all three axes move at the same time.
15. On the machines without the helical interpolation feature, a helix can be cut if the machine has the _______________, known as the A or B axis.

Relating the Concepts

No answers are suggested.

1. When programming a canned cycle on a machining center, how would you make sure that the tool moves freely over the clamps and bolts when changing hole positions?
2. How would you calculate drilling depth?
3. How would you calculate tapping depth?
4. List and explain two ways of setting up tapered tools for hole machining.
5. Explain how to program drilling deep holes.
6. List the causes for tap breakage when tapping.
7. Give examples and explain how to program the feed rate when tapping.
8. Give examples and explain how to program the automatic tool change.
9. What is the purpose of geometry offsets on a machining center?
10. Explain how to program a helix on machines that do not have the helical milling feature.

9 Macro Programming

Key Concepts

Subprograms and Macros
How Does a Macro Program Work?
Of What Does a Macro Consist?
- Arguments and Variables
- Mathematical Operators
- Mathematical Functions
- Control Instructions

Calling and Canceling a Macro Program
Strategies for Creating a Macro Program
Branching Examples
Looping Examples
Testing a Macro Program

Subprograms and Macros

A **macro** (short for *macroinstruction*) is a series of instructions that can be executed repeatedly by the machine control. A macro program enables a programmer to use the most advanced stage of automation on a CNC machine. Any programming task that can be performed on a CNC machine, from the simplest to the most complex, can be **automated** with a macro.

There are similarities and differences between a macro program and a subprogram. A macro program call is similar to a subprogram call that is repeated by address L. However, there is a difference in what is repeated. Whether a subprogram repeats a simple linear move, such as a cut-off operation, or a complete machining sequence programmed by a canned cycle, the repeated action is always the same or constant. A macro program can also repeat a simple or complex operation, but it can change one or more programmed values while the repetition is in progress. These changeable values are known as *parameters,* and the process of writing macros is known as **parametric programming.** (Note that parameters in a macro program differ from machine parameters.)

During the execution of macro programs, parameters vary according to the conditions set. This means that the use of variables is the main characteristic of a macro program. Consequently, macro programming is also defined as **variable programming.**

Another difference between a subprogram and a macro is that dimensional information in a subprogram is generally in incremental mode, while in a macro program it can be specified in either incremental or absolute mode.

It is not possible to list all programming tasks for which macros may be used. In general, macro programs are used when programming complex part shapes and when designing programs for a family of parts, but the application of macro programming depends only on the programmer's imagination. Almost any CNC department can benefit from adopting macro programming in order to improve efficiency.

How Does a Macro Program Work?

The part illustrated in Figure 9–1 can be machined using a macro program. On this part, there is a shoulder marked A that frequently has a different length. This dimension can be programmed in general form using a macro instruction.

The main program is in conventional form from beginning to end. At one point in the program, usually when the tool comes into position for cutting, a macro program is called up. After the macro program is executed, the control returns to the main program:

```
O10; (Main program number.)
N10 G50 X15.0 Z3.0 S1000 M42; (Coordinate system preset.)
N20 G00 T100; (Tool selection.)
N30 G96 S550 M03; (Spindle start.)
N40 G00 X2.0 Z0.1 T0101; (Rapid to position for straight cutting. Offset call.)
N50 G65 P901 A3.0; (Unconditional user macro program call.)
N60 G01 X3.1; (Shoulder cutting.)
N70 G00 X15.0 Z3.0 T0 M09; (Return to the home position. Offset cancel.)
M01; (Optional program stop.)
```

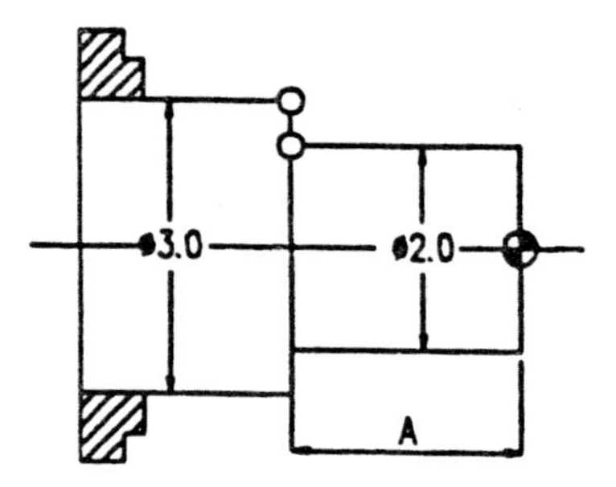

Figure 9–1 A simple macro program.

O901; (Macro program number.)
G01 Z-[#1] F0.01 M08; (Macro instruction to cut any shoulder length.)
M99; (Return to the main program.)

In this program, the tool is brought into position for cutting the 2.0 inch diameter. Then in line N50, the macro program is called by the G65 instruction and address P, which branches the execution to the macro program labeled O901. The word A3.0 represents the shoulder length that sometimes has to be changed. In the macro program, A is assigned to variable 1.

The next command cuts the shoulder length. This is performed by the following macro instruction:

G01 Z-[#1] F0.01;

To use a conventional command to cut a 3.0 inch shoulder length, one would enter:

G01 Z-3.0 F0.01;

However, this would allow cutting only this particular length. The macro command allows the programmer to cut any shoulder length desired. The particular shoulder length that is to be cut depends on the value set for address A in the **macro program call** (the G65 block). In this program the value for address A is 3.0, which means that a shoulder of this length will be cut.

After the length of the shoulder is cut, the control returns to the main program to cut a shoulder diameter. Indeed, if a shoulder diameter is to be changed according to a different part size, it may also be programmed using the macro command. For this purpose the address B will be entered in the main program. Then the program is changed as follows:

O10; (Main program number.)
N10 G50 X15.0 Z3.0 S1000 M42; (Coordinate system preset.)
N20 G00 T100; (Tool selection.)
N30 G96 S550 M03; (Spindle start.)
N40 G00 X2.0 Z0.1 T0101; (Rapid to position for straight cutting. Offset call.)
N50 G65 P901 A3.0 B3.1; (The address B added.)
N70 G00 X15.0 Z3.0 T0 M09; (Return home. Offset cancel.)
M01; (Optional program stop.)

```
O901; (Macro program number.)
G01 Z-[#1] F0.01 M08; (Macro instruction to cut any shoulder length.)
X[#2]; (Added macro instruction to cut any shoulder diameter.)
M99; (Return to the main program.)
```

Macro instructions allow programming in general form, adding a great deal of flexibility to the production system. Before machining a new part size, the operator can easily enter the main program and change the address values in the G65 block.

Normally, more sizes of this part, such as the diameter of the tool start point and/or the chamfer length, may be programmed in general form, if necessary. The more part sizes are programmed in general (macro) form, the more flexible the program is. For example, say one company makes several sizes of the same type of valve. Using conventional programming, ten programs may be needed to machine ten different sizes of valves. But there are several different parts in one valve. This means that many more programs would have to be created. In the long run it takes a lot of time and money. Macro programming requires only one program for each part of the valve. A minor change in the main program is needed to start machining on one particular part, since the macro program is designed in general form. It is quick and easy for the operator to make such a change in the G65 block by modifying the values for A, B, or any other address that may be used. Also, the program is already proven and less time is needed to check and confirm it. Thus, the first part can be produced faster, resulting in savings in programming and operating time.

Of What Does a Macro Consist?

There are some additional codes used to design the macro body. In the simple macro program just presented, the macro body consists of one command: G01 Z-[#1]. It is obvious that this is the instruction for straight cutting along the Z axis. Thus, no special consideration should be given to the macro body. However, sometimes the design of the macro needs more attention in order to produce mathematical equations using variables that describe the tool path in general form. These variables can be substituted for any suitable numbers, referred to as the *constants*. This enables programming various unusual part shapes in short programs that would otherwise be difficult to program because of the many calculations needed. Following these equations, the machine control performs all the necessary calculations using built-in software to get a properly machined part shape.

Arguments and Variables

An **argument** is a constant that must be provided to a macro program for the function to work. The **argument list,** which is placed in a user macro call statement (G65 block), consists of the values that will be sent to a macro program for processing. These arguments cannot be used directly in a user macro body. They have to be converted into variables or parameters according to a **variable list** of one particular machine. Following is a variable list taken from one popular control. (Keep in mind that your control may have a different list that you will have to follow when selecting arguments and variables for programming macros.)

A #1	B #2	C #3	D #7	E #8	F #9	H #11	I #4
J #5	K #6	M #13	Q #17	R #18	S #19	T #20	U #21
V #22	W #23	X #24	Y #25	Z #26			

On the preceding list, the argument is described by the letter, while each variable has a pound sign (#) in order to enable the control to distinguish it from ordinary numbers.

Usually, the arguments are the constants, or in other words, any suitable numbers assigned to the addresses presented on a variable list. They decide the starting value for each variable.

The arguments are placed in a call statement of a main program, while the variables are placed in a **macro body.** The relation between them is based on the variable list, which is specific for each machine. In order to better understand the relation between arguments and variables, consider the following programming block:

```
G65 P901 A1.0 B1.75 D3.0;
```

G65 is an unconditional user macro call that causes the following variables to be generated within a user macro body to cut, say, an ellipse.

```
#1 = 1.0 (Ellipse minor axis.)
#2 = 1.75 (Ellipse major axis.)
#7 = 3.0 (Stepping angle.)
```

When reading the G65 block, the control memorizes the values for A, B, and D, and relates these values to the corresponding variables #1, #2, and #7. Since the variables are in general form, the control is able to change the initial values during the program execution. For instance, a 3.0-degree stepping angle will change after every step-execution of the program as follows: 0, 3, 6, 9, 12, etc., until an angle set by the conditional expression is reached; i.e., 90.

If two arguments are assigned to the same variable, the last assignment is valid, following the same rule as for conventional instructions. For example, in the following macro call:

```
G65 D3 D10;
```

D10 is assigned to variable #7 (not D3) to express the ellipse's stepping angle.

Mathematical Operators

There is no difference between ordinary mathematical operators and those used in macro instructions.

+ Addition
– Subtraction
= Equals
• Multiplication
/ Division
, Comma, to separate words
() Brackets, to separate groups of operations and for comments
[] Square brackets, to define a variable or a constant

Mathematical Functions

These are mathematical functions used when creating macros. Most CNC programmers are already familiar with them:

sin	[...]	Sine (Unit: degree)
cos	[...]	Cosine (Unit: degree)
tan	[...]	Tangent (Unit: degree)
atan	[...]	Arc tangent (Unit: degree)
sqr	[...]	Square
sqrt	[...]	Square root
abs	[...]	Absolute value
bin	[...]	Binary (Conversion from binary to decimal.)
bcd	[...]	Conversion to binary
rnd	[...]	Round to an integer
fix	[...]	Round to a lower integer
fup	[...]	Round to a higher integer

Control Instructions

There are special programming instructions, known as **control instructions,** used to control the execution of a macro program. Machine tool builders have borrowed them from high-level computer programming languages such as BASIC and FORTRAN. This text reviews the most frequently programmed control instructions used to create macro programs on CNC machines.

At one point in a program, the programmer might need to branch the control execution. There are two types of **branching instructions** used in macro programming: conditional and unconditional branches. **GOTO** is an example of an **unconditional branch instruction.** In the following command:

```
GOTO 70
```

there is no condition set and the program execution is branched to line 70 at once. It works the same as calling a subprogram, and in fact, the GOTO branch instruction does not distinguish a macro program from a subprogram.

The **conditional branch instruction** distinguishes a macro program from a subprogram. Following is an **IF conditional branch instruction** used in conjunction with the GOTO instruction:

IF [conditional expression] GOTO N

This command executes a single instruction or group of instructions, depending on the outcome of a specified condition. If the **conditional expression** is true, one set of instructions will be executed. If the condition proves to be false, a second set of instructions will be executed. In either case, after the appropriate instructions have been executed, the control returns to the next sequential instruction.

There are a number of specific conditional branch instructions. Some of them branch on the value of a condition set; the others branch depending on the value of an ongoing count. Conditional expression can be one of the following mathematical expressions, where A and B addresses are used to express any values that might be entered:

[A] EQ [B]	Equal to
[A] NE [B]	Not equal to
[A] GT [B]	Greater than
[A] LT [B]	Less than
[A] GE [B]	Greater than or equal to
[A] LE [B]	Less than or equal to

The first two conditional expressions (equal to and not equal to) are the simpler, since the testing is performed in order to answer only Yes or No. Usually there is no calculation performed when testing these expressions. The last four expressions are more sophisticated, because a comparison is made (sometimes more than once) in order to check the relation between two values to see if a condition is satisfied. Usually, calculations are repeated before the testing is over. Consider the following conditional branch instruction:

IF [A] EQ [B] GOTO 60

If A (which is a variable, machine parameter, or constant) equals B (which is a variable, machine parameter, or constant), then the condition is true and program execution branches to line 60 by the statement GOTO 60. If A is not equal to B, then the condition is not true and program execution is continued in the following block. This simple macro feature enables the programmer to design one short program for two different jobs, or in principle to perform two different actions in the program. A block related to a GOTO statement must have a sequence number; otherwise, the control will not find the block and will display an alarm message.

Many programming applications are concerned with executing the same sequence over and over again. This is referred to as a **program loop.** The process of executing the same sequence of instructions or loop over and over again is known as **looping.** This can be accomplished by programming the **DO loop** control instruction.

Looping allows programmers to achieve an iteration (repetition) in a macro program. It is one of the most valuable macro programming techniques. But it should be remembered that any time a program loop is set up, the programmer must make certain that there is a way out of the loop; otherwise, a control execution will stay inside the loop indefinitely.

The objective of a DO loop is to execute the instructions the number of times determined by control parameters specified by the DO and WHILE statements. The repetition of a DO loop is performed by the **WHILE conditional expression** that is placed at the beginning of the loop. The instructions to be repeated are programmed inside the loop body. These are variables, usually in the form of mathematical equations. How many times the DO loop is repeated, or how the execution is performed until the operation is completed, depends on the initial conditions set. These conditions cannot be changed once the loop is activated. In order to finish and exit the loop, the loop must contain an **END statement.** The following illustrates the structure of the DO loop:

```
WHILE [Conditional expression] DO m
... (Body, which consists of mathematical equations and move commands.)
END m
```

The *m* represents the number of times that the loop is repeated. The number in the END statement must match the number in the DO statement.

In the example WHILE [A LE 45] DO1, the control may be instructed to perform mathematical equations and move commands. While A is less than or equal to 45, the condition is true and the DO loop is repeated. When the END statement is reached, the control is returned to the WHILE statement for new conditional testing. If the condition is still true, a new step execution is performed by the DO statement. When A is greater than 45, the conditional expression is not satisfied and execution of the block following the END instruction is performed. Usually the control is then returned to the main program.

The process of looping involves the following steps:

1. Execute the DO statement and activate the DO loop.
2. Compute the iteration count.
3. Test the iteration count.
4. Execute the statements in the loop body.
5. End the loop according to the condition set.

Since the iteration count is set initially, the control parameters inside the loop cannot be changed during the execution of the DO loop. Also, control variables cannot be changed once the loop execution is started. The DO loop becomes inactive when the iteration count is finished. The looping is over when the control reaches the END statement.

More than one WHILE control instruction can be used in the same DO loop of the macro program. This is known as *nesting.* This means that inside the DO loop there can be more than one control statement, but they have to be written in a certain order so that each END instruction matches the most recent WHILE instruction, as shown at the top of the next page:

```
WHILE [B=2] DO1 ─────────────┐
    WHILE [A=3] DO1 ───────┐ │
    END ───────────────────┘ │
END ─────────────────────────┘
```

The inside loop is executed first. When it ends, the outside loop is executed, and complete looping is over. By nesting the loops, it is possible to solve the most complicated programming tasks in very short programs.

Calling and Canceling a Macro Program

On some popular controls the macro program is called by the G65 code. A macro program number is found by assigning the P address in the same way as when calling a subprogram. Program execution is returned to the main program using the M99 command. When iteration is to be programmed, a macro program must end with the END statement. Then, more instructions may be added if desired before calling the M99 and returning to the main program. Following is an example of calling and canceling the macro program:

```
G65 P901 A1.0 B1.5 C2.0 E3.0; (Unconditional macro call.)
M99; (Return to the main program.)
```

More than one call statement may be placed in the main program, but each must have its macro program with the M99 return command, which returns the control to the line after the G65 macro call. On some controls a macro program can be called from a subprogram, and a macro program can call a subprogram in conventional or macro form.

Strategies for Creating a Macro Program

A programmer usually programs a user macro call after he or she sets the coordinate system, makes the tool change, and turns on the spindle. The programmer should study the blueprint, tooling, and machining operations in detail to decide which values should be assigned as variables in a macro program. The values of any address that is to change during the execution of the program have to be programmed as variable. The most common mistake programmers make in designing macros is not assigning one of these addresses as a variable or not expressing it properly by a mathematical equation.

It is very important that the tool start point be set properly according to the assigned variables. Generally speaking, the tool should be positioned at its start point before activating the DO loop in a user macro body, but the repeated move command should be placed inside the DO loop. Then program execution continues inside the loop until the condition is satisfied. Since the program control does not go back to reread the information before the loop for executing each step, it can process the information and execute the program faster. The creation of macro programs is discussed in detail in the following examples.

Branching Examples

The following practical problem is a good example of the use of branching. There are times when different bar stock sizes are used to machine the same part. The G71 canned cycle is normally used for roughing. If the program is designed for a 3.0-inch-diameter bar stock, the cycle start point may be programmed as:

```
G00 X3.0 Z0.1;
```

When a larger bar stock is used, say 3.25 inch, it may be programmed as:

```
G00 X3.25 Z0.1;
```

it is possible to use the G71 canned cycle for the different bar stock sizes (or predrill sizes when boring), since the part finish profile is the same. This means that all programmed data remains the same; only the cycle start point changes. The control calculates the number of passes necessary to get the job done.

To avoid an operator mistake when changing the program, a safer and more efficient solution is to use a macro program with a conditional branch instruction. To program this, the programmer can label the bar stock size using the address S (as stock), B (as bar), or D (as diameter). Actually, it is up to the programmer to describe and give a logical meaning to each argument and variable pair in the macro program. He or she should always do this according to the variable list of the particular machine being used. (We will choose the address B to avoid possible confusion with speed or degree if S or D are selected.)

The programmer sets a true or false condition using the numbers 1 and 0 as the flags. When a 3.0 inch diameter is cut, the operator sets a flag to 1, producing a true state. The tool is moved to the start position by the G00 X3.0 Z0.1; instruction. When machining the part from 3.25 inch bar stock, the condition is false (the flag is set at 0). The program branches to move the tool to a corresponding start point programmed by the G00 X3.25 Z0.1; instruction. These instructions, which position the tool to the different canned cycle start points, are placed in a macro body. The following program illustrates the use of branching instructions:

```
O10; (Main program.)
G50 X15.0 3.0 S1500 M42; (Coordinate system preset.)
G00 T0303; (Tool and tool offset call.)
G96 S510 M03; (Spindle start.)
G65 P902 B1; (User macro call. Flag set to 1.)
```

```
G71...; (Canned cycle.)
...; (Machining.)
G00 X15.0 Z3.0 T0 M08; (Return home.)
M01; (Optional program stop.)

O902; (Macro program number.)
N30 IF [#2 EQ 1] GOTO 60; (Condition with B in G65 block. Note: #2
    corresponds with B in G65 block.)
N40 G00 X3.25 Z0.1; (Rapid to position for cutting a 3.25 inch diameter.)
N50 GOTO 70; (To jump over N60.)
N60 G00 X3.0 Z0.1; (Rapid to position for cutting a 3.0 inch diameter.)
N70 M99; (Return to the main program to start a canned cycle.)
```

When machining the part from 3.0 inch bar stock, the operator keeps the flag as 1. When machining from 3.25 inch bar stock, he or she changes the flag to 0. Then branching begins: Line N30 instructs the control to jump to N60 to start machining a 3.0-inch-diameter bar stock because the condition is true [#2 EQ 1]. If the condition is false, the execution continues in the N40 block to start machining a 3.25 inch diameter. In this case, the block N50 is important since the control should jump to line N70, preventing line N60 from being read in. In either case, when line N70 is read, the control returns to the main program by the M99 code and execution continues by the programmed canned cycle.

The presented program combines the conditional and unconditional branching in the N30 and N50 blocks, respectively. There are several variations of setting the conditional expression for this task, but all produce the same result.

Looping Examples

A program used to cut an ellipse on the lathe is one example of looping (Figure 9–2). There is no machine function to cut an ellipse directly to final shape. Therefore, the G01 function must be programmed to make a series of linear motions for a specified angle, which will be changed according to a conditional expression set initially.

Usually, Z0 is set on the part front face, but in this case it should be set at the intersection of the major and minor ellipse axes. Then it is obvious that the loop execution starts from 0 degrees and finishes when the count is 90 degrees. The WHILE control instruction may be used to perform the repetition of the loop for a specified stepping angle.

An argument in the G65 block will contain a 5-degree stepping angle. This argument refers to degree, so it is practical to use the address D, which corresponds with variable #7 from the variable list presented earlier. For each step execution, the angle increases 5 degrees, as set by the address D. After the angle changes from 0 to 90 degrees, the ellipse is complete. While an ellipse is being cut, the coordinates of the tool's position change. Consequently, the X and Z coordinates have to be expressed as variables by mathematical equations (Figure 9–3).

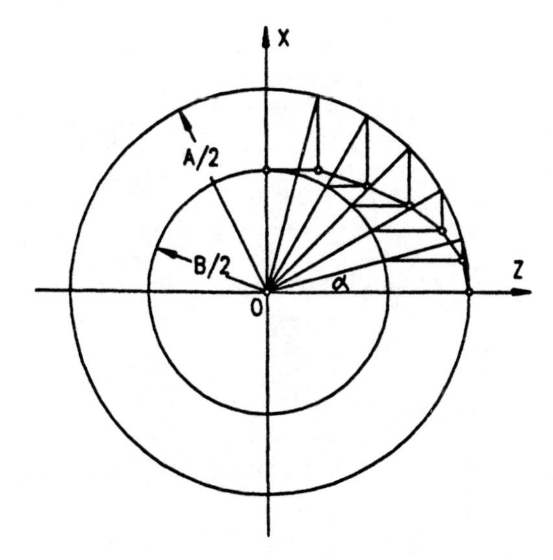

Figure 9–2 An ellipse cut with multiple G01 function calls.

This must be set in advance; once looping starts, there is no help if the part shape is not described properly. The program might work without error, but the result would be nonconformance of the part.

Following are expressions for the X and Z variables according to the triangles in Figure 9–3:

X value: (B / 2) • sin α
Z value: (A / 2) • cos α

On the Z axis, the tool will start from 0 degrees and finish when the count is 90 degrees. Therefore, the following WHILE instruction should be used:

```
WHILE [#7 LE 90] DO1
```

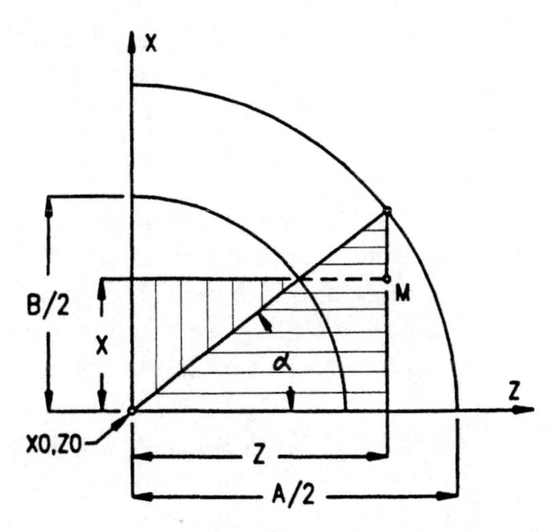

Figure 9–3 Mathematical equations for X and Z variables.

Variable #7 corresponds to the address D, which describes a stepping angle. After expressing the variables, the programmer can write a complete program to cut an ellipse using equations for the X and Z values. They will be placed inside the DO loop of the macro program as follows:

```
O301; (Main program.)
G50 X12.0 Z3.0 S1100 M42; (Coordinate system preset.)
G00 T0303; (Tool and tool offset call.)
G96 S550 M03; (Spindle start.)
G65 P505 A0.75 B1.875 D5.0; (Unconditional user macro call. The values for
    A, B, and D will be passed to variables in a user macro body.)
G00 X12.0 Z3.0 T0300 M09; (Return home. Offset cancel.)
M01; (Optional program stop.)

O505; (Macro program number.)
#101 = #1 / 2; (Minor axis; Half of the minor axis (B / 2) is assigned to a vari-
    able named #101.)
#102 = #2 / 2; (Major axis; Half of the major axis (A / 2) is assigned to a vari-
    able named #102.)
#103 = #102; (Used to move the tool to touch the face, which is on a distance of
    A / 2 from the part origin.)
G00 X0 Z[#103 + 0.2]; (The tool is at the part centerline on the X axis and 0.2
    inch from the part face on the Z axis.)
G01 Z[#103] F0.015 M08; (Feeding to touch the part face. Coolant on.)
#104 = #7; (Counter; #7 is the variable that counts a number of degrees.)
WHILE [#104 LE 90] DO1; (Condition; The execution is performed once by the
    DO1 statement and repeated until variable #104 is less than or equal to 90.)
#110 = #101 • SIN [#104]; (Equation for the X value. It is the same as: B / 2 •
    sin α, calculated earlier.)
#120 = #102 • COS [#104]; (Equation for the Z value. It is the same as: A / 2 •
    cos α, calculated earlier.)
G01 X[2 • #110] Z[#120] F0.007; (Move command according to equations for
    the X and Z values.)
#104 = #104 + #7; (Counter; This increases the count when a new step execution
    is performed. The increase is made for every 5 degrees, as set by D5.0. It is
    the same as: D + D + D ... etc. or 5 + 5 + 5 ... etc. until it reaches 90.)
END; (The loop must end with the END statement.)
M99; (Return to the main program.)
```

One variable can be a result of arithmetic operations between the other variables. This is called a *variable substitution.* As seen in this program, the control is able to remember the values for variables #101 and #102, and also to perform arithmetic operations. This means that they do not have to be reentered each time they are needed. This helps to write shorter programs. As a general note, once a macro program is finished and the M99 code is read in, these variables are forgotten and their registers are set vacant, having no value.

A program to mill an internal hemisphere, as shown in Figure 9–4, shows how to use looping when programming machining centers. It is assumed that excess material has been removed by predrilling. The hemisphere will be finished using a ball end mill with a suitable radius. In this case, an end mill with a 0.25 inch radius will do well.

The G01 function will be used to cut the line after each step execution set by the WHILE statement and the D address. On the Z axis, the tool will start from 0 degrees and finish when the count is 90 degrees. Therefore, the following WHILE instruction will be used:

```
WHILE [#7 LE 90] DO1
```

Variable #7 corresponds to the address D, which describes a stepping angle. A 360-degree circle will be programmed in the X–Y plane starting from, and returning to, Y0. To describe a full circle, it is enough to program the I value; the X and Y coordinates do not have to be entered. The tool path in the X–Z plane can be described using the X and Z values from the triangle in Figure 9–4. Thus, the I, X, and Z must be expressed as variables, as shown below.

$X = Rp \cdot \cos \alpha$ (Tool position on the X axis when the tool is feeding in.)
$Z = Rp \cdot \sin \alpha$ (Tool position on the Z axis when the tool is feeding in.)
$I = Rp \cdot \cos \alpha$ (Arc center displacement value when cutting a full arc.)

To use the same program for the family of parts, the radius of the hemisphere as well as the radius of the end mill have to be expressed as variables too. This is done in a macro body as follows:

```
O304; (Main program.)
G00 G90 G54 X0 Y0; (Coordinate system preset. Rapid to the part center.)
G43 H3 Z1.0 S900 M03; (Spindle start. Tool offset call.)
```

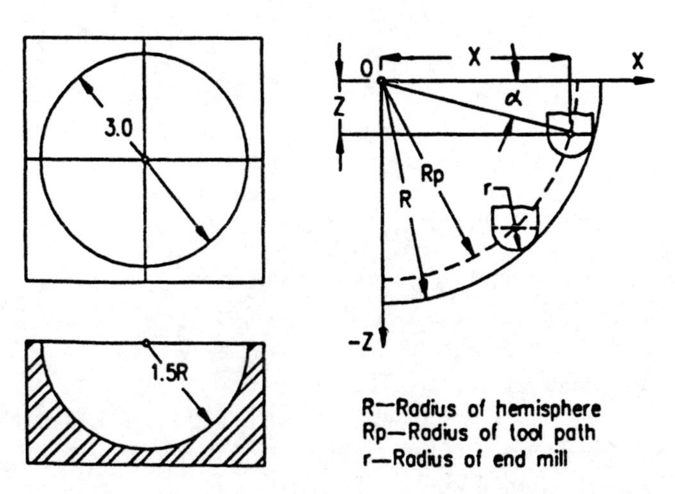

Figure 9–4 Milling an internal hemisphere.

```
Z0.1 M08; (Closer to the part on the X axis. Coolant on.)
G65 P912 A1.5 B0.25 D5.0; (Macro program call.)
G00 Z1.0 M09; (Retract on the Z axis. Coolant stop.)
G28 Z0; (Return home on the Z axis.)
M01; (Optional program stop.)

O912; (Macro program number.)
#101 = #1; (Radius of hemisphere.)
#102 = #2; (Radius of the ball end mill.)
#103 = #1 - #2; (Radius of the tool path.)
#104 = #7; (Counter that counts the number of degrees.)
G00 X[#103]; (Positioning on the X axis.)
G01 Z0 F6.0; (Feeding to touch the surface.)
WHILE [#104 LE 90] DO1; (Condition to be tested.)
#110 = #103 • COS [#104]; (X value.)
#120 = #103 • SIN [#104]; (Z value.)
G01 X[#110 Z – [#120] F3.5; (Feeding on X and Z.)
G02 I-[#110]; (Cutting a 360-degree circle.)
#104 = #104 + #7; (Increasing the counter.)
END; (Ending the loop.)
M99; (Return to the main program.)
```

Flexibility is one of the most valuable characteristics of macro programming and that gives it a distinct advantage over conventional programming. For instance, if the surface finish is not satisfactory, the D value in the main program should be changed. Instead of 5 degrees, a 3- or 2-degree angle could be used. If one macro program is used for the family of parts, then the A value (radius of hemisphere) and the B value (radius of ball end mill) should be changed according to specifications. Obviously, conventional programming does not have such capabilities.

Testing a Macro Program

If a macro program does not do what it is supposed to, or if an incorrect instruction results in an error message, the programmer must debug the macro, checking it for errors in syntax and errors in logic. If a logic error exists, it usually can be found in the mathematical equations entered to express the part shape. The programmer should also check the DO loop. The control will help him or her find a syntax error by generating an alarm message in each block where an error exists.

Following are some suggestions that might be of help in developing and running macro programs:

1. Plan the macro. Figure out what you want to accomplish before you start writing it.
2. Make a list of the main steps for creating a macro for one particular task. Name the variables and draw the sketch.

3. Test the macro. Play the program mentally, going through each instruction trying to predict the outcome when it is executed.
4. Use the "On-screen Simulation" feature of the CAM software, if available.
5. Run the macro program with care, using machine safety features.

Although different controls might have slightly different rules for creating macro programs, they do not differ in essence. They all use the same computer logic, which allows the programmer to get into the higher stage of automation of CNC equipment.

Summary

A macro is a series of instructions to the control that can be executed repeatedly. A macro program call is similar to a subprogram call that is repeated by address L. Macros can repeat simple or complex operations, changing one or more programmed values while the repetition is in progress. These changeable values are known as the parameters; because of that, programming a macro is known as parametric programming. This is also defined as variable programming, since the variable is a main characteristic of a macro program.

Branching instructions branch the control execution at one point in the program. Two types of branching instructions are used in macro programming: conditional and unconditional. Using an unconditional branch is the same as calling a subprogram—the program branches to the line specified without any condition. The conditional branch works depending on the outcome of the specified condition. If the conditional expression is true, one set of instructions will be executed; if the condition proves to be false, a second set of instructions will be executed. In either case, after the appropriate instructions have been executed, the control returns to the next sequential instruction.

Looping, one of the most valuable macro programming techniques, is used to achieve iteration in the program. The objective of a DO loop is to execute the instructions the number of specified times determined by control parameters in the DO and WHILE statements. The repetition of a DO loop is performed by the WHILE conditional expression, which is placed on the beginning of the loop. These conditions cannot be changed once the loop is activated. In order to finish and exit the loop, an END statement has to be programmed. Since the iteration count is set initially, the control parameters inside the loop cannot be changed during the execution of the DO loop. Also, control variables cannot be changed once the loop execution is started. The DO loop becomes inactive when the iteration count is finished. The looping is over when the control reaches the END statement.

Programs with macros are short and efficient. They are mostly used when programming for a family of parts and various unusual part shapes that otherwise require lengthy programs and part programming software for calculating tool positions. Because of the specifics of contour milling, macros are more often used on machining centers than on lathes.

Although different controls might have slightly different rules to follow when creating macro programs, they do not differ in essence. All controls use the same computer logic, which allows getting into the most advanced stage of automation on CNC machines.

Key Terms

argument
argument list
automated
branching instructions
conditional branch instruction
conditional expression
control instructions
DO loop
END statement
GOTO
IF conditional branch
looping
macro
macro body
macro program call
parametric programming
program loop
unconditional branch instruction
variable list
variable programming
WHILE conditional expression

Self-Test

The answers are in Appendix E.

1. A _______________ is a series of instructions that can be executed repeatedly by the machine control.
2. Any programming task that can be performed on a CNC machine can be _______________ with a macro.
3. Because of changeable values known as parameters, the macro is called _______________.
4. Variable is the main characteristic of a macro program; consequently, macro programming is also defined as _______________.
5. A _______________ is programmed using the G65 code.
6. The design of _______________ needs attention in order to produce mathematical equations using variables that describe the tool path in general form.
7. An _______________ refers to the constant that must be provided to a macro program for the function to work.
8. The _______________ is placed in a user macro call statement (G65 block).

9. Arguments must be converted into variables according to a _______________ of one particular machine.
10. _______________ are special programming functions used to control the execution of a macro program.
11. The _______________ branch the program execution.
12. An _______________ works the same as calling a subprogram since there is no condition set when branching.
13. The _______________ statement is an example of an unconditional branch instruction.
14. A _______________ branches the program execution depending on the outcome of the specified condition.
15. The _______________ instruction is used in conjunction with the GOTO statement.
16. Mathematical expressions such as "equal to" or "greater than" are the _______________ used in macro programming.
17. Through the use of a _______________, the same sequence can be executed over and over.
18. The iteration in a macro program is achieved using the _______________ technique.
19. The objective of a _______________ is to execute the instructions a number of times determined by control parameters specified by the DO and WHILE statements.
20. The repetition of a DO loop is performed by the _______________.
21. In order to finish looping and exit the loop, the _______________ must be programmed.

Relating the Concepts

No answers are suggested.

1. Explain the difference between a subprogram and a macro program.
2. List the advantages of macro programming over conventional programming.
3. Explain the use of the G65 block.
4. Recognize and define a variable list.
5. Define and explain unconditional branching instructions.
6. Define and explain conditional branching instructions.
7. Recognize and define the meaning of variable substitution.
8. Define and explain looping instruction.
9. Recognize and describe a WHILE control instruction.
10. Explain a nesting technique in macro programming.

10 Using Computers in CNC Programming

Key Concepts

The Machine Control Computer
- Control Modes
- Operations in Memory Mode

Personal Computers
- Software
- Secondary Storage

Personal Computers and Part Programming
- The CNC Editor
- Part Programming Software

Conversational Programming
- The APT Programming Language
- COMPACT II

The Machine Control Computer

The computer on numerically controlled machines controls the program execution; hence this computer is known as the *machine control* or *controller* (Figure 10–1). This computer performs tasks similar to a personal computer, such as loading the program, searching through the program, altering and deleting information, and executing the program (program run).

Figure 10–1 A CNC vertical machining center equipped with a modern control that utilizes 32-bit logic for on-screen programming, foreground/background operations, maintenance and standard functions for program editing, and program execution. (Courtesy Hitachi Seiki U.S.A., Inc.)

The machine control computer consists of two main components: **hardware,** which is any physical component in a computer system, such as the processor, monitor, memory, and input/output devices; and **software,** which is a general term for programs stored in the memory in electronic form.

The CNC programs are stored in computer memory, which also holds the software needed to run the control computer. To change and alter data, the programmer and the operator need to write into and read from memory. The main memory of a computer is **RAM (Random Access Memory).** RAM is volatile memory; the programmer and the operator can easily change the content of this memory by writing over the old programs and replacing them with new ones. On a CNC machine control computer, this is where the programmer and the operator store the CNC programs and read and write as they like. When the power to the machine is turned off, the content of this memory is lost. To save data in memory when the machine is turned off, the machine uses its batteries. The machine switches to batteries automatically. When the batteries weaken, a light on the control panel lights up, warning the operator to change the batteries. This should be done within a month, or the data stored in RAM will be lost permanently.

Besides RAM, the computer has **ROM (Read Only Memory).** This memory allows reading from it, but writing to it is not normally possible. The computer must have ROM in order to store some important information, which is system, not user, related. Since the user cannot write into ROM, he or she cannot change its content. The content of ROM does not change when the power is turned off. Various CNC software (such as canned cycles) is stored in this memory for permanent use.

Control Modes

There are two control modes of the machine control computer; memory mode and manual mode. Memory mode is used for memory operations, and manual mode is used for manual operations. No manual operation is possible in memory mode, and no memory operation is possible in manual mode.

The memory mode is divided into three main functions: Memory, Edit, and MDI (Manual Data Input). Manual mode is also divided into three functions: Handle, Rapid, and Zero Return. The operator may select a particular function using the mode selector on the operator panel (Figure 10–2).

The **Memory function** is used to load the program from an available source such as floppy disk, perforated tape, or communication channel; to punch out the program from the machine; and to run/execute the program. Program editing cannot be performed using this function. One can see the program, but one cannot edit it by saving the changes. If a programmer tries to edit the program in Memory mode, most controls will display an alarm message. This is a good safety feature, since it reminds the programmer that he or she is trying to edit using the wrong procedures. The programmer must use the CYCLE START button to start program run.

Any changes and alterations entered in the program will be saved through the **Edit function.** This function works in conjunction with the editing keys such as RESET, SEARCH, INSERT, DELETE, and ALTER. In the Edit mode, the following operations are possible:

- Loading a program from an office computer via a communication channel or loading from a tape
- Resetting the program to the beginning

Figure 10–2 Mode selector.

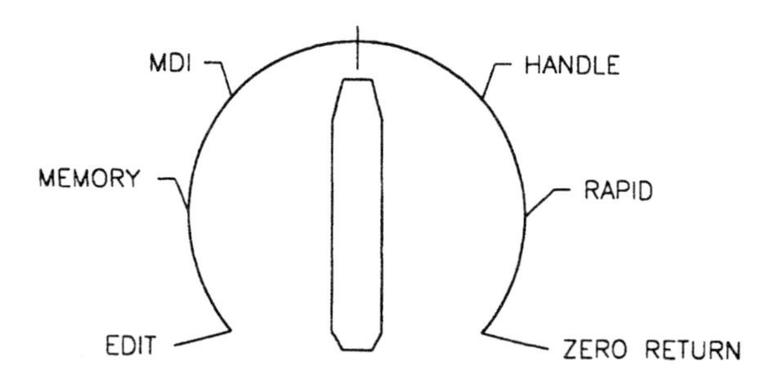

- Searching for a particular piece of data or another program in memory
- Tape editing by alteration, deletion, or insertion of data
- Deleting a complete program from memory
- Program punch-in using the machine keyboard
- Program punch-out to an office computer, or tape punch-out
- Parameter setting
- Diagnostics and maintenance

The **MDI function** is used for manual input of information, such as tool offset, and for setting machine parameters and diagnostics. Any information entered with MDI will be saved by the control computer, except when using MDI for programming. This means that a programmer can also use MDI to make a small program or to enter a single command, but after it is executed, the control clears the command from memory. On some controls it still may be written on the MDI page, but it cannot be executed again unless the programmer rewrites the command. To execute a command entered in MDI, the programmer and the operator use the START button or the CYCLE START button, depending on the machine/control combination.

The operator must use the manual commands Handle and Rapid to move the machine slides from one position to another during such manual operations as the setup of tools. The Zero Return function allows the operator to return the machine to its origin when he or she needs to confirm the distance from that fixed point. When the slide reaches the machine origin, the light comes on. When moving the machine slide close to the machine travel limits, the control may generate an alarm if the operator does not use the Zero Return function.

Operations in Memory Mode

All of the controls use appropriate keys to allow the operator to manipulate the program in memory. These keys have the same functions on all the machines and are labeled similarly.

On some occasions, such as an interruption in program execution or after program editing, the operator needs to reset the program to rewind it to the beginning. On most controls, when the program is reset in Memory mode, it stops the program execution, the spindle rotation, and the coolant, but it does not return the program control to the beginning of the program. When the operator pushes the CYCLE START button, the program execution starts from where it stopped, which can cause serious problems later. In Edit mode, the program control always returns to the head of the program when reset. When the operator pushes the CYCLE START button, the program execution starts from the beginning of the program. Good operators get in the habit of resetting the program in Edit mode.

When an operator searches for a particular piece of information in the memory, the control may be in either Memory or Edit mode. The operator can search through the screen pages, or he or she can use the **SEARCH key.** For example, the operator punches in the block number, tool number, tool offset number, or any other particular address for which he or she is searching, and presses the SEARCH key. The cursor will stop under the requested information. On older controls, a search runs from the beginning to the end of a program, but newer controls allow the operator to search up and down with help from the appropriate arrow keys.

To change information in a program, the operator selects the Edit mode and calls the block with a particular piece of information to be changed. He or she then places the cursor under the information to be changed and punches in a new piece of information, which is shown at the bottom of the screen. The operator then presses the **ALTER key** and checks that the change has been made. Pressing the RESET button then resets the program.

To delete the information, the operator selects the Edit mode and calls the block with the information to be erased. He or she places the cursor under the particular piece of information to be deleted and presses the **DELETE key.** The operator then checks that the information has been deleted.

To insert information, the operator selects the Edit mode and calls the block where the particular piece of information is to be inserted. He or she places the cursor under the information where the new information is to be inserted and types in a new piece of information, which will be shown at the bottom of the screen. The operator presses the **INSERT key** and checks that the change has been made.

Personal Computers

Personal computers are being used more and more for creating, editing, and storing the programs for the CNC machines. They have almost replaced tape-punching facilities because of their obvious advantages of speed, readability, dependability, and large storage capacity.

These computers can be simple units or workstations with special programming software. They are used in a number of ways in the creation of part programs, such as:

- Text editing and database creation
- Simulating tool movements
- Running programming software
- Graphical numerical control

Personal computers are usually connected to the CNC machines via communication channels. The program can be sent to the machine or received from it. All newer CNC machines have this interface capability for transmitting and receiving data. Different machines have different transmission speeds, which are called the *Baud rate.* When several CNC machines are linked on an office computer, this speed is set according to the machine with the lowest Baud rate.

Software

The software runs the computer. It also enables the user to use the computer for different applications. Only a specific task can be accomplished through the use of a specific software. There are two types of software; application software and system software.

System software, the operating system for one or a group of computers, connects the hardware and application software in a system that inputs, processes, and outputs data (Figure 10–3). The most widely accepted operating systems for personal computers used in CNC applications are DOS (Disk Operating System) and Windows.

In order to execute an application, an operating system must perform a number of support functions, such as program loading and data copying. The user communicates with the operating system through the command language. Typically, there is a simple, one-word command for each function that the operating system must perform. Some of these commands are: FORMAT, COPY, DISKCOPY, DELETE, and LOAD. These commands can be entered by typing them on the keyboard (DOS) or selecting a graphic icon using a mouse (Windows).

Application software processes data. Different applications have different purposes. For example, you can purchase part programming software, or software to create a spreadsheet, a database, or a document (word processing).

Secondary Storage

No matter how large it is, RAM cannot store all the data that one user may need to store. The extension of main memory is used for this purpose, and it is known as **secondary storage.** The most common computer secondary storage medium is a floppy disk, also called a *diskette.*

There are two sizes of floppy disks, $5\frac{1}{4}$ inches and $3\frac{1}{2}$ inches. Floppy disks are also distinguished by capacity (how much data each can hold). The most common

Figure 10–3 The operating system connects the hardware and application software in a system for processing data.

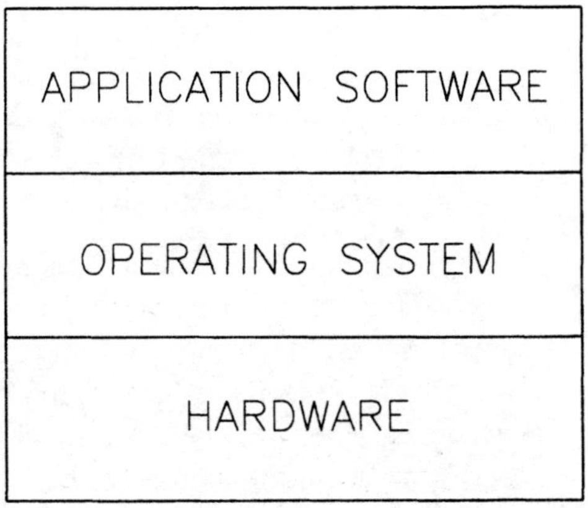

APPLICATION SOFTWARE
OPERATING SYSTEM
HARDWARE

types are standard-capacity floppy disks (also called *double-density floppy disks*) and high-capacity floppy disks (also called *high-density floppy disks*). The capacity of floppy disks is measured in bytes (B), kilobytes (KB), and megabytes (MB) of data. One byte is equal to one character (a letter, number, or special character such as $, %, or #). One KB is equal to approximately one thousand bytes, and one megabyte is equal to approximately one million bytes. The following table summarizes diskette characteristics and common labeling notation:

SIZE	CAPACITY	COMMON LABELLING NOTATION
5¼″	360KB	5¼″ Double-Sided Double-Density (5¼ DSDD)
5¼″	1.2MB	5¼″ Double-Sided High-Density (5¼ DSHD)
3½″	720KB	3½″ Double-Sided Double-Density (3½ 2DD)
3½″	1.44MB	3½″ Double-Sided High-Density (3½ 2HD)

The floppy disk drive, or diskette drive, is the device that reads and writes data on the floppy disk. The drive is usually accessed through a slot on the processing unit, but some drives are stand-alone units. In general, a standard floppy disk drive can read and write to only standard-capacity floppy disks, while a high-density disk drive can read and write to either standard or high-density floppy disks.

The PC usually has one or two floppy disk drives, known as drive A and drive B. Some PCs have both 5¼-inch and 3½-inch high-density drives.

Many common technical applications, such as drafting software, are so large that they need a large disk space. Time required to retrieve information from a large floppy disk (or a number of disks) is noticeably long because the processing unit takes so long to rotate the disk in the drive and access the selected information. A **hard disk** spins constantly at a faster speed than diskette drives; therefore, data on a hard disk can be accessed much faster than data on a floppy disk.

Another advantage of a hard disk, usually labeled drive C or drive D, is its storage capacity. A typical 5¼-inch double-sided floppy disk can hold 360,000 characters; most 3½-inch disks can hold about 1.4 MB of data; while hard disks, even on PCs, can store hundreds of megabytes of data.

Personal Computers and Part Programming

A **CNC program file** is the data for a given application grouped together. A programmer usually creates the file using the word processing software or using the text editor that comes with almost all application software. A file may contain a few lines or a large number of pages. A CNC program file is usually from one to a few pages long, depending on the length of the CNC program.

When using a word processor in CNC programming, the program is treated as text. After the program is created, it can be sent to the machine via the communication channel and run. These word processors use the ASCII code set, which is the same as the CNC machine code set. For longer programs, this saves time over punching in the programs via the machine computer keyboard.

Programs can be saved on a hard disk in a database directory (a library of files) for later use. If there is no communication link between a personal computer and the machine, a tape puncher can be connected to the computer to produce the punched tape from which the program is loaded into the machine and run.

CNC Editor

There are special text editors on the market called CNC editors. As the name implies, they are especially designed to help CNC programmers in their jobs. One CNC editor may have the following functions:

- Editing the program
- Resequencing the program (to assign sequence numbers to the N address)
- Outputting ASCII or ISO programs to the CNC machine
- Inputting ASCII or ISO programs from the CNC machine
- Calling operating system functions: delete, copy, rename, print, etc.

The Edit function usually has some special assignments, which can be invoked by the touch of keyboard function keys. These are:

- Editing the current program
- External editing (editing any program in the current directory while working on one particular program)
- Copying part of the current program to another place in the program
- External copying (copying a part of any program in the current directory to a program presently being worked on)
- Moving one part of the program to another place
- Deleting a part of the program

These functions enable the programmer to create several exemplar programs, such as for drilling, turning, boring, threading, and so on. When the same operation has to be programmed, the programmer can copy a particular exemplar program to a current program using the External Copy function. All of the necessary changes, such as tool numbers and the cycle start point, are done quickly and the program is ready to go, speeding up the programming.

Remember that a computer without part programming software can only be used to enter the final information. Any necessary calculations have to be performed by the programmer manually or in any other convenient way. In this case, the computer is only a tool to type the program and to send it to the machine if there is a communication link.

Part Programming Software

Part programming software is used to ease programming for CNC machines when a complex part geometry requires calculation of a large number of tool positions. Part programming software is usually incorporated into a family of CAM

(Computer Aided Manufacturing) software. Some CAM software is associated with CAD (Computer Aided Design) software into CAD/CAM stations. Then the CAM software can use the CAD files as a source of data, which speeds up the programming process.

Part programming software is user-friendly, meaning the programmer does not have to know the computer programming language or its operating system. It uses screen menus to lead the user through the programming process. Data can be entered via the keyboard, the mouse, the tablet, or the function keys. Experienced programmers can use built-in macro capabilities and advanced techniques such as a family of parts to become even more productive.

Programming software has a dynamic graphics database to hold the actual machining sequences. These sequences can be viewed, edited, chained, or deleted. The programming can be accomplished whether single cuts or CNC machine canned cycles will be used. The software will also automatically calculate the proper feeds and speeds to be used during the machining, create a tooling list, and define the tool path.

Programmers can use different layers to associate with each profile being created or to construct clamps and fixtures to get a complete picture of the part setup. The tool motion can be seen as it will occur at the machine (Figure 10–4).

Using part programming software, the programmer can easily solve trigonometry problems to define an accurate tool path. When the program is done, the programmer can send it from the PC to the machine via a communication channel using built-in software with communications capability. Good part programming software is capable of:

1. Establishing the machining parameters and tooling for a particular machine or job
2. Defining the geometry and tool path
3. Code generation, enabling the programmer to define what code is to be generated and how it is output to the machines
4. Communication, enabling the programmer to use standard communications protocols or create his or her own

Some part programming software facilitates 3-D machining and projection onto a surface in 3-D. This is a powerful tool when engraving in molds. Surface finishing passes with a boll nose end mill called *pocketing* might be a problem for the programmer, but when using a good programming software, it is a simple task.

Conversational Programming

There are several computer programming languages that developers use to create CNC programming software. They are conversational, logical in their approach, and user-friendly. The programmer uses English-like conversational statements or

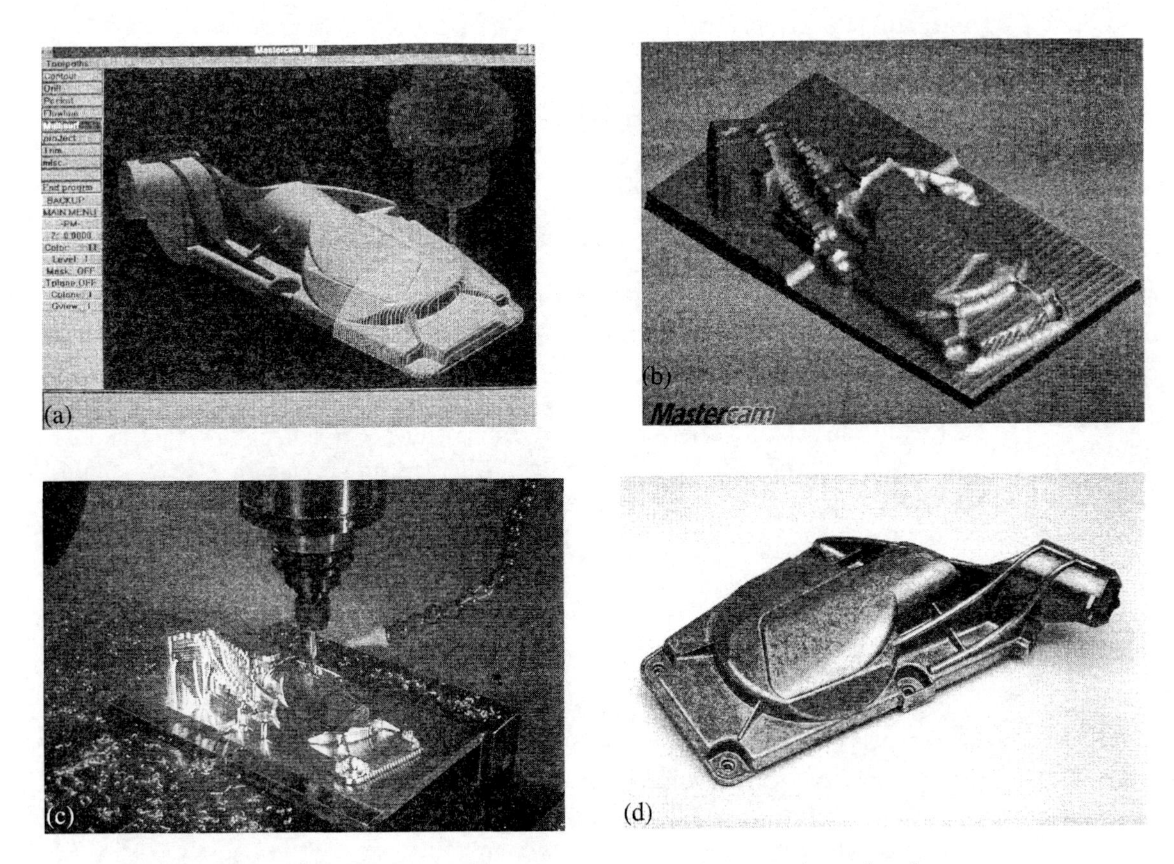

Figure 10–4 CAD/CAM software in action. The part shown is a car mirror chassis. The chassis was designed and programmed in Mastercam (slide a). The tool path was run through Mastercam's solids-based verification utility, giving a realistic model of the cut part (slide b). A prototype was cut (slide c) and approved. Finally, the part went into production (slide d). This series is an excellent example of what an inexpensive Windows-based CAD/CAM package is capable of. (Courtesy CNC Software, Inc.)

source code to select the machine and tools and to describe a tool path. The programming software accepts these statements, makes calculations if needed, and then translates a source code into appropriate code.

Once the program is written, the programmer needs an output file in order to use it on the CNC machine. To enable the CNC machine control to read and execute the prepared program, the output file must be compatible with the tape format of one particular machine, including the G and M codes. However, in the CNC field there are variations in tape format from one machine/control combination to another. The answer is a postprocessor.

The **postprocessor** is a set of instructions used to transform tool centerline data into machine-tool motions in a computer numerically controlled system. The postprocessor produces an output file, including feed rates, spindle speeds, and auxiliary functions, that may be printed, punched, or downloaded directly to the machine control via a communication channel.

Two of the more common general programming languages for numerical control programming are APT and COMPACT II. They both process information in four phases:

1. Reading input files and scanning for errors
2. Performing arithmetic operations to define a tool path on a two- or three-dimensional surface
3. Editing to apply coordinates of a tool path to different machining sequences
4. Postprocessing; producing NC codes acceptable to the particular machine/control combination

Phases 1, 2, and 3 describe the tool path using source code (English-like language) through APT or COMPACT II statements. Phase 4 (postprocessing) translates source code into NC codes of a particular machine using a MACHIN statement. The postprocessing is accomplished through the use of the postprocessor software, which customizes the output and produces the necessary syntax format for a particular machine. Usually, the postprocessor is not included in the APT or COMPACT II package and must be bought at additional cost from the machine tool builder, the control system builder, or the software developer.

The APT Programming Language

The **APT (Automatically Programmed Tool)** programming language was developed in the early 1960s to assist engineers in defining the proper instructions and calculations for NC part programming. Originally, APT was a programming language for a large mainframe computer used in large companies such as the McDonnell Douglas Corporation. Today, a full-feature APT can be used on minicomputers, which makes it a popular choice of programming language worldwide, standardized under ANSI (American National Standards Institute) and ISO. Also, its two-dimensional version ADAPT can be used on a desktop computer. There are several versions of APT in use today, but they do not differ significantly.

A great strength of APT is its ability to perform precise calculations for complicated tool paths when contouring on a three-dimensional surface in a multiaxis programming mode. Because of this, APT is more in use for programming on machining centers than on lathes; the shapes of the lathe parts do not match the complexity of some milled parts. Also, APT makes programming too complicated when applied to the lathe parts with simple two-axis geometry. For lathe programming, programmers mostly use two-dimensional programming languages or CAD/CAM software and canned cycles.

APT part programming is based on the concept that two reference surfaces, called the *part* and the *drive,* guide the tool toward a third reference surface called a *check surface* (Figure 10–5).

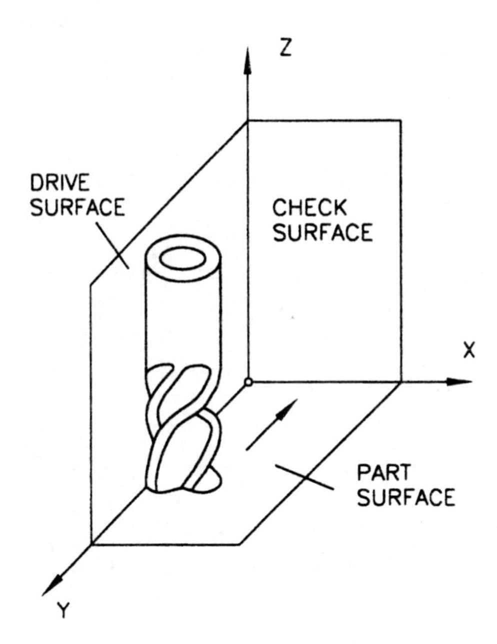

Figure 10–5 APT programming concept.

Figure 10–5 shows that when the tool is cutting along the Y axis using the X-Y (G17) plane as the part surface, the Y-Z plane is the drive surface and the X-Z plane is the check surface. The situation changes when the tool is cutting along the X axis using the X-Y plane as the part surface: The X-Z plane becomes the drive surface and the Y-Z plane becomes the check surface. Thus, to guide the tool in different cutting directions, APT must switch the reference surfaces as needed.

The tool may be positioned on, to the left, or to the right of the drive surface while cutting. According to Figure 10–6, the tool may be programmed to stop TO, ON, or PAST the check surface.

When the tool stops at the check surface, programmed either by TO, ON, or PAST, the tool may go right by GORGT, left by GOLFT, forward by GOFWD, backward by GOBACK, up by GOUP, and down by GODOWN. Some of these commands are illustrated in Figure 10–7. Note that these are tool motions from the last position. In addition, the programmer can instruct the control to move the tool to a specific point using a GOTO statement. For incremental programming on either axis, the programmer uses a GODLTA statement.

Before moving the tool around the part, the programmer must describe the part geometry using the geometry statements. He or she must describe various entities such as points, lines, and circles using the letter P for point, L for line, and C for circle. The letter is followed by the number, and they are separated by the slash character. Any combination of letters and numbers may be used to describe the entity, provided the total does not exceed six characters, and at least one of them must be a letter. The geometry statement to define a point in space is

P1 = POINT/X coordinate, Y coordinate, Z coordinate

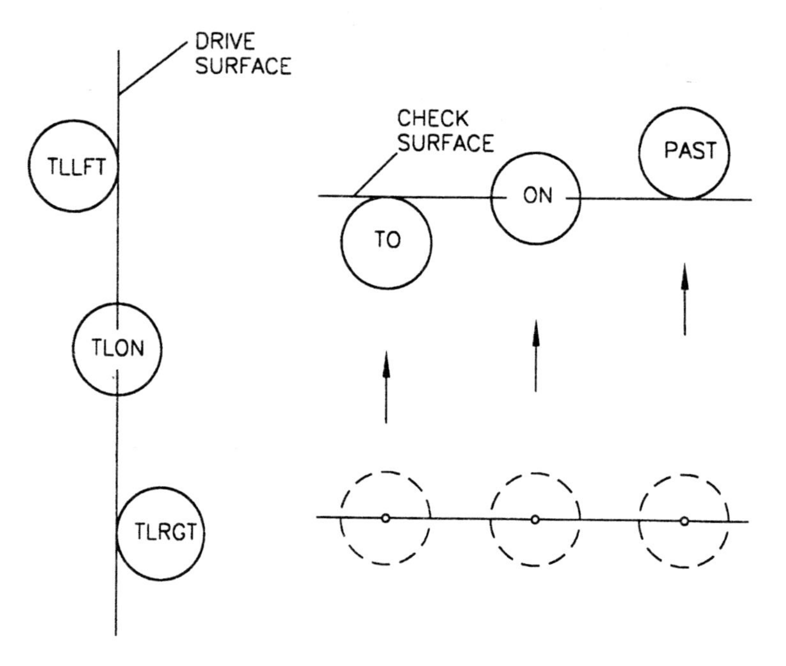

Figure 10–6 Tool position in relation to check and drive surfaces.

Figure 10–8 shows some geometry statements (note that APT uses many more geometry statements, not shown here due to space limitations). The particular point in 3-D is described as:

P1 = POINT/5,-2,3

Following is a part of a typical APT program and translated NC code to drill a ½-inch-diameter hole, 0.75-inch deep at position X5.0, Y2.0, Z0, using a GOTO statement.

Source Code:

```
MACHIN/MILL,48NCV (Machine selection.)
PRINT/ON (Drawing setup.)
CALLPRN (Drawing setup.)
SETPT=POINT/0,0,1.0 (Defining the tool start point.)
PT1=POINT/5.0,2.0 (Defining point 1.)
LOADTL/2 (Loading Tool 2.)
CUTTER/0.5 (Tool diameter.)
SPINDL/450,RPM (Spindle on at 450 RPM.)
COOLNT/ON (Coolant on.)
RAPID,FROM/SETPT (Rapid traverse from reference position.)
CYCLE/DRILL,0.75,0.005,IPR,0.1 (Drilling cycle specified to drill 0.75 inch
    deep, using 0.005 inch feed per spindle revolution, with 0.1 inch R value.)
GOTO/PT1 (The tool moves to point 1 to start machining.)
CYCLE/OFF (Cycle end.)
```

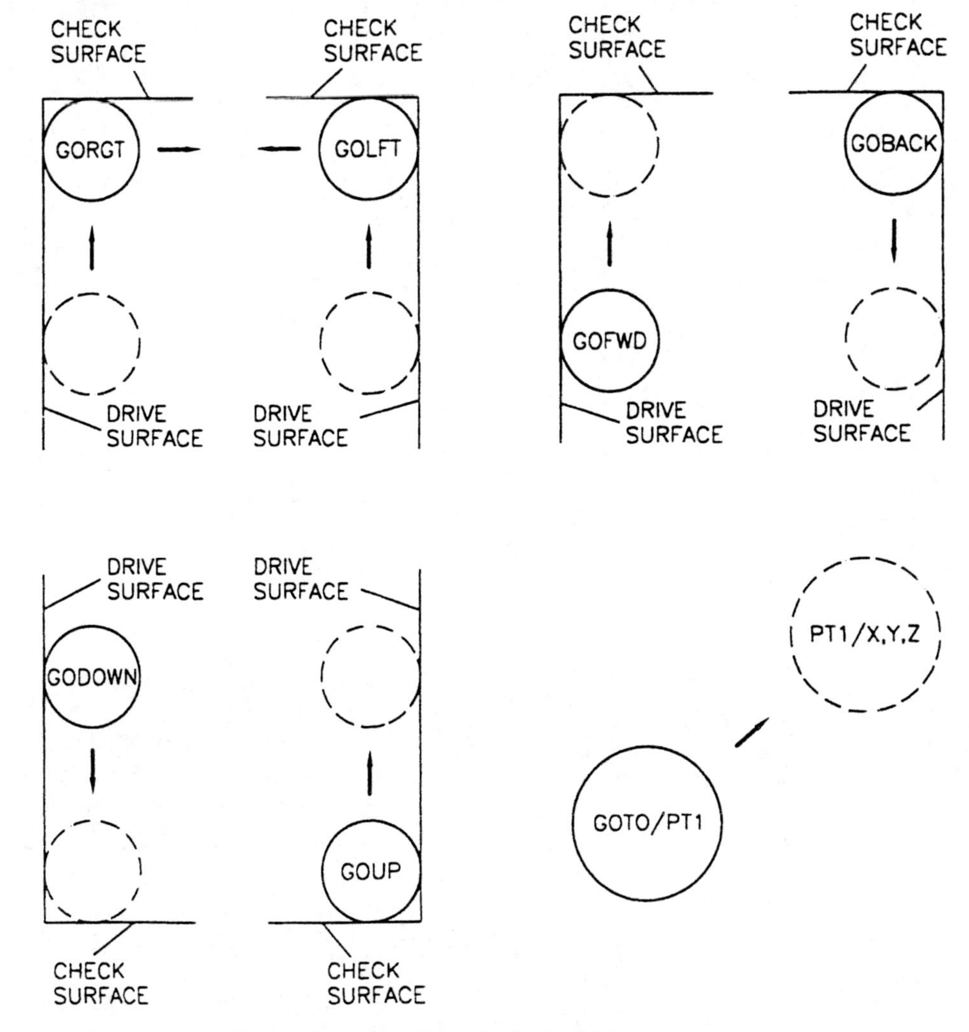

Figure 10–7 Possible tool motions from the last position.

RAPID,GOTO/SETPT (Return to tool start point.)
END (End of sequence.)
FINI (End of program.)

NC Code:

N10 T2 M6; (Tool change.)
N20 S450 M03; (Spindle start at 450 RPM.)
N30 M08; (Coolant on.)
N40 G81 X5.0 Y2.0 Z-0.75 F0.005 R0.1; (Drilling cycle.)
N50 G80; (Canned cycle cancel.)
N60 G0 Z1.0; (Retract to the tool start position on Z.)
N70 X0,Y0; (Retract to the tool start position X and Y.)

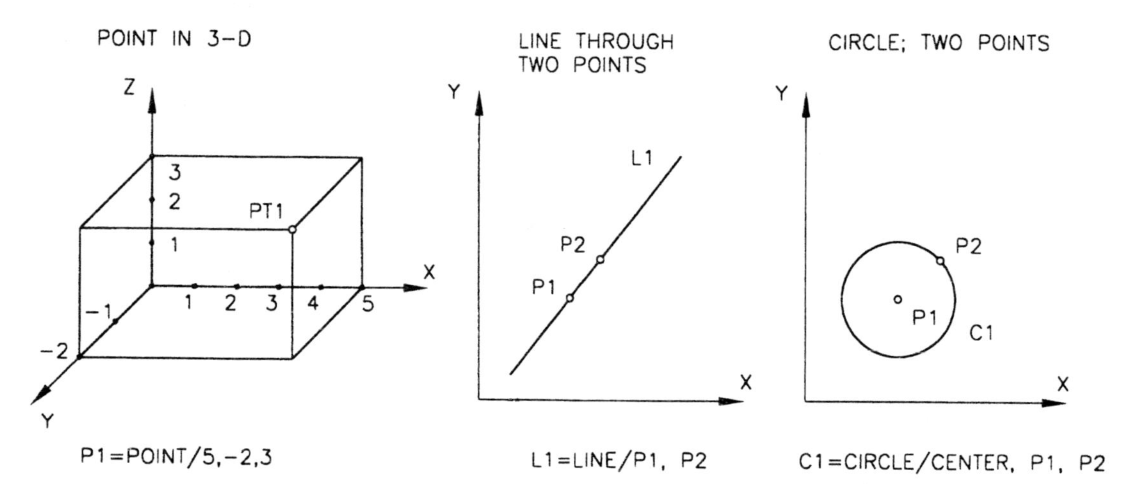

Figure 10–8 Some APT geometry statements.

N80 M01; (Optional program stop.)
M90 M30; (Program end.)

In this program the emphasis is not to produce a useful part; rather it is on illustrating which APT source codes produce their counterparts in CNC code.

COMPACT II

The popularity of the **COMPACT II** programming language is based on two facts: From its inception it was designed to be used on personal computers, and it is easy to learn. The COMPACT II programming concept is similar to APT programming, offering simple rules and English-like statements. This programming language uses reserved symbols for defining entities, such as DPT for point, DLN for line, and DCIR for circle. The programmer uses a MOVE statement to move the tool in rapid traverse and a CUT statement to move the tool by the feed rate. To move the tool in cutting mode TO, ON, or PAST the check surface, the programmer uses CUT,TOLN, CUT,ONLN, and CUT,PASTLN, respectively. COMPACT II also uses GOTO statements to move the tool to a particular position. Following is a typical COMPACT II program and translated NC code to machine the part in Figure 10–9.

Source Code:

MACHIN,MILL9006 (Machine identification.)
IDENT,H2400 ADAPTER (Part identification.)
SETUP,LX,15LY,LZ (Drawing setup.)
DRAW,-18.OXA,-1OYA,SCALE.25 (Scaling the drawing.)
;TOOL/ON;GEOM/ON (Tool and geometry on.)
BASE,XA,YA,ZA (Establishing the drawing reference point.)

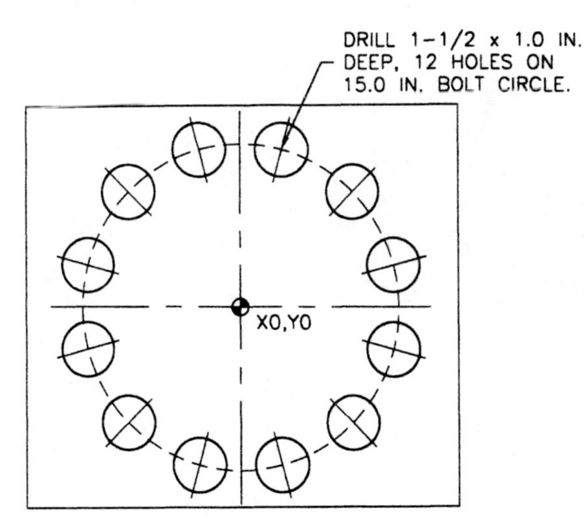

Figure 10–9 COMPACT II programming example.

DRAW,PEN8,AXIS (Drawing specifications; pen 8.)
DLN1,YB (Line parallel to Y axis.)
DLN2,XB (Line parallel to X axis.)
DPT100,LN1,LN2,ZB (Part origin.)
DSET1,PT100,15.5BC,-15CW,12EQSP,NOMORE (Spacing bolt circle; the software calculates the coordinates for hole locations.)

$ CENTERDRILL

DRAW,PEN2 (Using pen 2.)
ATTCHG,TOOL10,TLCMP1,30FPM,0.005IPR,TD0.125,180TPA (Tool selection. Tool clamp. Speed and feed rate specification.)
SENSE/ON (Sense on.)
DRL,SET1,0.020DP,0.1 CLEAR (Drilling cycle specified as 1.0 inch initial point level, 0.02 inch depth, and 0.1 R point level.)
END (End of sequence.)

$ 1-1/2″DRILL

DRAW,PEN9 (Using pen 9.)
ATCHG,TOOL30,TLCMP2,60FPM,0.008IPR,TD1.5,130TPA (Tool selection. Tool clamp. Speed and feed rate specification.)
SENSE/ON (Sense on.)
DRL,SET1,1.2DP,0.1CLEAR (Drilling cycle specified as 1.0 inch initial point level, 1.2 inch depth, and 0.1 R point level.)
END (End of sequence.)
FINI (Program end.)

CNC Code:

N100; (Center drill.)
N001 G00 G90 X0 Y15.0 E0.005 S917 T10 H01 M03 M08;
N002 G81 X7.4859 Y2.0059 Z-0.02 K.25;
N003 Y-2.0059;

```
N004 X5.4801 Y-5.4801;
N005 X2.0059 Y-7.4859;
N006 X-2.0059;
N007 X-5.4801 Y-5.4801;
N008 X-7.4859 Y-2.0059;
N009 Y2.0059;
N010 X-5.481 Y5.4801;
N011 X-2.0059 Y7.4859;
N012 X2.0059;
N013 X5.4801 Y5.4801;
N014 G80;
N015 M01;

N200; (1-1/2 drill.)
N020 G00 G90 X0 Y-1.773 E0.008 S153 T30 H02 M03 M08;
N021 G81 X7.4859 Y2.0059 Z-1.2 K.25;
N022 Y-2.0059;
N023 X5.4801 Y-5.4801;
N024 X2.0059 Y-7.4859;
N025 X-2.0059;
N026 X-5.4301 Y-5.4801;
N027 X-7.4859 Y-2.0059;
N028 Y2.0059;
N029 X-5.4801 Y5.4801;
N030 X-2.0059 Y7.4859;
N031 X2.0059;
N032 X5.4801 Y5.4801;
N033 G80;
N034 M01;
N035 M30;
```

This program is made for a machine with the Zero Shift function programmed by the G90 code. The programmer does not have to enter the coordinate system preset code in the program. Before entering a new machining sequence, the machine first returns to the home position set by the operator. Another characteristic of this machine/control combination is sensing capabilities; the control senses when the tool touches the part surface and reacts by establishing the part zero on the Z axis. Because of this, the programmer enters SENSE/ON for drilling operations or SENSE/OFF for other operations in the COMPACT II source program. Consequently, for drilling operations, the operator does not have to calculate the tool length offsets.

Summary

The computer on the CNC machine controls the program execution; hence this computer is known as the machine control. It consists of two main components: hardware, which is any physical component in a computer system, such as the processor, monitor, memory, and input/output devices; and software, which is a general term for programs stored in the memory in electronic form.

The main memory of a computer is RAM (Random Access Memory). RAM is volatile memory; the programmer and the operator can easily change the content of this memory by writing over the old programs and replacing them with new ones. On a CNC machine control computer, this is where the programmer and the operator store the CNC programs and read and write as they like. When the power to the machine is turned off, the content of this memory is lost. To save data in memory when the machine is turned off, the machine uses its batteries.

The control computer also has ROM (Read Only Memory). This memory allows reading from it, but writing to it is not normally possible. The content of ROM does not change when the power is turned off. Various CNC software (such as canned cycles) is stored in this memory for permanent use.

There are two control modes of the machine control computer: memory mode and manual mode. Memory mode is used for memory operations, and manual mode is used for manual operations. Manual operations are not possible in memory mode, and memory operations are not possible in manual mode.

The memory mode is divided into three main functions: Memory, Edit, and MDI (Manual Data Input). Manual mode is also divided into three functions: Handle, Rapid, and Zero Return.

The Memory function is used to load the program from an available source such as floppy disk, perforated tape, or communication channel; to punch out the program from the machine; and to run/execute the program. Program editing cannot be done using this function. The Edit mode is used to change, alter, or delete the program. This function works in conjunction with the editing keys such as RESET, SEARCH, INSERT, DELETE, and ALTER. The MDI function is used for manual input of information such as tool offset and for setting machine parameters and diagnostics. Any information entered through MDI will be saved by the control computer, except when using MDI for programming purposes.

Personal computers are being used more and more for creating, editing, and storing the programs for CNC machines. They have almost replaced tape-punching facilities because of their obvious advantages of speed, readability, dependability, and large storage capacity. These computers can be simple units, or workstations with special programming software. They are used in a number of ways in the creation of part programs, such as:

- Text editing and database creation
- Simulating tool movements
- Running programming software
- Graphical numerical control

Personal computers are usually connected to CNC machines via communication channels. The program can be sent to the machine or received from it. All newer CNC machines have this interface capability for transmitting and receiving data.

The software runs the computer. It also enables the user to use the computer for different applications. Only a specific task can be accomplished through the use of a specific software. There are two types of software: application software and system software.

System software, the operating system for one or a group of computers, connects the hardware and application software in a system that inputs, processes, and outputs data. In order to execute an application, an operating system must perform a number of support functions, such as program loading and data copying. The user communicates with the operating system through the command language: a simple, one-word command for each function that the operating system must perform.

Application software processes data. Different applications have different purposes, such as part programming, or creating a spreadsheet, a database, or a document (word processing).

RAM cannot store all the data that one user may need to store. The extension of main memory is used for this purpose, and it is known as secondary storage. The most common computer secondary storage medium is a floppy disk, also called a diskette. A hard disk is also used as secondary storage. A computer can access data on a hard disk faster than on a diskette. A hard disk has much larger storage capacity than a diskette.

A CNC program file is the data for a given application grouped together. A programmer usually creates a file using the word processing software or using the text editor that comes with almost all application software. A CNC program file is usually from one to a few pages long, depending on the length of the CNC program.

When using the word processor in CNC programming, the program is treated as text. After the program is created, it can be sent to the machine via the communication channel and run. These word processors use the ASCII code set, which is the same as the CNC machine code set. For longer programs, this saves time over punching in the programs via the machine computer keyboard.

There are special text editors on the market called CNC editors. They are especially designed to help CNC programmers in their jobs. One CNC editor may have the following functions:

- Editing the program
- Resequencing the program (to assign sequence numbers to the N address)
- Outputting ASCII or ISO programs to the CNC machine
- Inputting ASCII or ISO programs from the CNC machine
- Calling operating system functions: delete, copy, rename, print, etc.

A computer without part programming software can only be used to enter the final information. Any necessary calculations have to be performed by the programmer manually or in any other convenient way. In this case, the computer is only a tool to type the program and to send it to the machine if there is a communication link.

Part programming software is used to ease programming for CNC machines when a complex part geometry requires calculation of a large number of tool positions. Part programming software is usually incorporated into a family of CAM (Computer Aided Manufacturing) software. Some CAM software is associated with CAD (Computer Aided Design) software into CAD/CAM stations. Then the CAM software can use the CAD files as a source of data, which speeds up the programming process. Using part programming software, the programmer can easily solve trigonometry problems to define an accurate tool path. When the program is done, he or she can send it from the PC to the machine via a communication channel using a built-in software with communications capability.

There are several computer programming languages that developers use to create CNC programming software. They are conversational, logical in their approach, and user-friendly. The programmer uses English-like conversational statements or source code to select the machine, tools, tool path, and so on. The programming software accepts these statements, makes arithmetic calculations if needed, and then translates a source code into appropriate code.

Once the program is written, the programmer needs an output file in order to use it on a CNC machine. To enable the CNC machine control to read and execute the prepared program, the output file must be compatible with the tape format of one particular machine, including the G and M codes. The output file is generated using a postprocessor, a set of instructions that transforms tool centerline data into machine-tool motions in a computer numerically controlled system. The postprocessor produces an output file, including feed rates, spindle speeds, and auxiliary functions, that may be printed, punched, or downloaded directly to the machine control via communication channel.

Two of the more common general programming languages for numerical control programming are APT and COMPACT II. A great strength of APT is its ability to perform precise calculations for complicated tool paths when contouring on a three-dimensional surface in a multiaxis programming mode. APT part programming is based on the concept that two surfaces, called the *part* and the *drive,* guide the tool toward a third surface called a *check* surface. The COMPACT II programming concept is similar to the APT programming concept, offering simple rules and English-like statements that make it easy to learn and use.

Key Terms

ALTER key
application software
APT
CNC program file
COMPACT II
DELETE key
Edit function
floppy disk
hard disk
hardware
INSERT key
MDI function

Memory function
part programming software
postprocessor
RAM (Random Access Memory)
ROM (Read Only Memory)
SEARCH key
secondary storage
software
system software

Self-Test

The answers are in Appendix E.

1. _______________ refers to any physical component of a computer.
2. _______________ is a general term for computer programs.
3. The _______________ is used for manual input of information.
4. The _______________ is used for program execution or program run.
5. When editing a program using the _______________, the control saves all of the changes entered.
6. The _______________ is volatile memory to which a user can read and write.
7. The _______________ is nonvolatile memory from which a user can read, but writing is not normally possible.
8. The _______________ is used to search for information in the program.
9. The _______________ is used to change information in the program.
10. The _______________ is used to delete information in the program.
11. The _______________ is used to insert information in the program.
12. _______________ is the extension of main memory.
13. The _______________ helps the user in processing data.
14. _______________ is the operating system for one or a group of computers.
15. A _______________ is the medium for recording data.
16. Data on a _______________ can be accessed much faster than data on diskette.
17. A _______________ is the data for a given application grouped together.
18. _______________ is used to ease programming.
19. _______________ are computer programming languages applied to CNC programming.
20. A _______________ is a set of instructions used to transform tool centerline data into machine-tool motions.

Relating the Concepts

No answers are suggested.

1. Explain the difference between the RAM and ROM.
2. Differentiate between resetting the program in the Edit mode and in the Memory mode.

3. Explain how to insert a piece of information into the memory of a CNC machine.
4. Explain how to delete a piece of information from the memory of a CNC machine.
5. List the advantages of the hard drive over a floppy disk.
6. Differentiate between system software and application software.
7. List the functions of the CNC editor.
8. Explain how exemplar programs can help in programming.
9. Explain the APT programming concept.
10. Explain the role of a postprocessor.

Appendix A
CNC Code List

G-Code List

G00	rapid, positioning
G01	linear interpolation
G02	circular interpolation, clockwise
G03	circular interpolation, counterclockwise
G04	dwell*
G10	variable data input
G11	variable data input cancel
G14	helical interpolation, clockwise
G15	helical interpolation, counterclockwise
G17	X-Y plane, machining center
G18	X-Z plane, machining center
G19	Y-Z plane, machining center
G20	inch data input
G21	metric data input
G22	activation of safety limits, lathe
G23	deactivation of safety limits, lathe
G27	automatic zero return, check*
G28	automatic zero return to the reference point*
G29	automatic return from the reference point*
G30	return to second reference point*
G32, G33	thread cutting with a constant lead, lathe
G34	thread cutting with an increased lead, lathe
G35	thread cutting with a decreased lead, lathe

*nonmodal

G40	tool radius compensation cancel
G41	tool radius compensation left
G42	tool radius compensation right
G43	tool length compensation in the same direction, machining center
G44	tool length compensation in the opposite direction, machining center
G50	coordinate system preset and maximum spindle RPM, lathe*
G52	local coordinate preset*
G53	machine coordinate selection*
G54, G55, G56, G57, G58, G59	work coordinates preset*
G70	finishing cycle, lathe*
G70	inch data input, some lathes
G71	roughing cycle, lathe*
G71	metric data input, some lathes
G72	face roughing cycle, lathe*
G73	pattern repeat cycle, lathe*
G74	peck grooving cycle on the Z axis, lathe*
G75	peck grooving cycle on the X axis, lathe*
G76	single-pass threading cycle, lathe*
G80	canned cycle cancel, machining center*
G81	drilling cycle, machining center*
G82	drilling cycle with dwell, machining center*
G83	peck drilling cycle, machining center*
G84	tapping cycle, machining center*
G85	boring cycle, machining center*
G86	alternate boring cycle, machining center*
G87	alternate boring cycle, machining center*
G88	alternate boring cycle, machining center*
G89	boring cycle with dwell, machining center*
G90	diameter cutting cycle, lathe*
G90	absolute programming, machining center
G91	incremental programming, machining center
G92	coordinate system preset, machining center*
G92	threading cycle, lathe*
G94	face cutting cycle, lathe*
G94	feed rate in inch per minute, some lathes
G95	feed rate in inch per revolution, some lathes
G96	constant surface speed control
G97	constant revolution per minute
G98	return to initial point level in fixed cycle, machining center
G99	return to R-point level in fixed cycle, machining center
G98	feed rate in inch per minute, lathe
G99	feed rate in inch per revolution, lathe

M Code List

M00	program temporary stop
M01	optional stop
M02	program end
M03	spindle start clockwise
M04	spindle start counterclockwise
M05	spindle stop
M06	tool change, machining center
M07	oil mist, machining center
M08	coolant on
M09	coolant stop
M10	chuck clamp, lathe
M10	fourth axis clamp, machining center
M11	chuck unclamp, lathe
M11	fourth axis unclamp, machining center
M12	tailstock spindle out, lathe
M13	tailstock spindle in, lathe
M16	tool post rotation, shortest direction, lathe
M17	tool post rotation, normal direction, lathe
M18	tool post rotation, reverse direction, lathe
M19	spindle orientation, machining center
M21	tailstock direction, forward, lathe
M22	tailstock direction, backward, lathe
M23	chamfering on, lathe
M24	chamfering off, lathe
M30	reset and rewind
M31	chuck bypass on, lathe
M31	interlock bypass on, machining center
M32	chuck bypass off, lathe
M32	interlock bypass off, machining center
M33	tool change, some machining centers
M48	override cancel on, machining center
M49	override cancel off, machining center
M41	spindle low speed range, lathe
M42	spindle high speed range, lathe
M50	oil hole coolant on, machining center
M51	air blow on, machining center
M54	high pressure coolant on, machining center
M56	high pressure coolant off, machining center
M73	Y axis mirror image off, machining center
M74	Y axis mirror image on, machining center
M75	X axis mirror image off, machining center
M76	X axis mirror image on, machining center
M73	parts catcher out, lathe
M74	parts catcher in, lathe

M85	automatic door open, lathe
M86	automatic door close, lathe
M98	subprogram call from a main program
M99	main program call from a subprogram

Address Description

A	a. threading tool tip angle
	b. indexing about the X axis, machining center
B	indexing about the Y axis, machining center
C	indexing about the Z axis, machining center
D	a. cutter radius compensation number, machining center
	b. cutting depth for turning and boring cycles, lathe
	c. number of divisions for pattern repeating cycle, lathe
	d. depth of first pass in threading cycle, lathe
E	a. feed rate function, machining center
	b. precision feed rate for threading, lathe
F	feed rate
G	preparatory function
H	a. tool length compensation number, machining center
	b. position compensation number
I	a. arc center modifier for the X axis
	b. taper height, X axis, lathe
	c. vector for cutter radius compensation for the X axis
J	a. arc center modifier for the Y axis, machining center
	b. shift amount for the Y axis in a fixed cycle, machining center
K	a. arc center modifier for the Z axis
	b. taper height for the Z axis, lathe
	c. vector for cutter radius compensation for the Z axis
	d. thread depth for repetitive threading cycle, lathe
L	a. subprogram repetition count
	b. fixed cycle repetition count, machining center
M	miscellaneous function
N	block number
O	program number (EIA)
P	a. subprogram number call
	b. macro program number call
	c. offset number
	d. start block number for turning and boring cycle, lathe
	e. dwell
Q	a. end block number for turning and boring cycle, lathe
	b. depth of each cut in peck drilling cycle, machining center
	c. shift amount in boring cycle, machining center
R	a. arc radius designation
	b. retract point in fixed cycle, machining center

S	spindle speed function
T	tool function
U	a. incremental dimensioning for the X axis, lathe
	b. stock allowance on the X axis in repetitive cycles, lathe
	c. dwell time
W	a. incremental dimensioning for the Z axis, lathe
	b. stock allowance on the Z axis in repetitive cycles, lathe
X	X axis absolute coordinate value designation
Z	Z axis absolute coordinate value designation

Appendix B
Handy Formulas for the CNC Programmer and Operator

Calculating the RPM (inch units):

$$RPM = \frac{FPM \cdot 3.82}{D}$$

RPM revolutions/minute
FPM cutting speed in feet/minute
D diameter of work for turning or diameter of tool for milling in inches

Calculating FPM (inch units):

$$FPM = \frac{RPM \cdot D}{3.82}$$

RPM revolutions/minute
FPM cutting speed in feet/minute
D diameter of work for turning or diameter of tool for milling in inches

Calculating the RPM (metric units):

$$N = \frac{1000 \cdot V}{D \cdot \pi}$$

N spindle speed in revolutions/minute
V velocity or cutting speed in meters/minute
D diameter of work for turning or diameter of tool for milling in mm
π constant = 3.14

Calculating cutting speed (metric units):

$$V = \frac{D \cdot \pi \cdot N}{1000}$$

N	spindle speed in revolutions/minute
V	velocity or cutting speed in meters/minute
D	diameter of work for turning or diameter of tool for milling in mm
π	constant = 3.14

Calculating feed rate (inch units):

Turning and drilling: F = RPM • IPR
Milling by end mill: F = RPM • IPT • NT

IPT	feed in inches/tooth
RPM	revolutions/minute
IPR	feed in inches/revolution
NT	number of teeth

Calculating feed rate (metric units):

Turning and drilling: F = RPM • IPR
Milling by end mill: F = RPM • FPT • NT

FPT	feed in millimeters/tooth
RPM	revolutions/minute
IPR	feed in millimeters/revolution
NT	number of teeth

Thread Data:

$$\text{Pitch} = \frac{1}{\text{Number of threads/inch}}$$

Calculating root diameter:

OD – 2 • S (external thread)
OD + 2 • S (internal thread)
S Single depth (height of thread)

Finding bore size when OD of a male thread is known:

$$\text{DIA bore} = \text{OD} - 2 \cdot \text{S}$$

Finding depth of thread:

UN thread	$H = 0.61343 \cdot P$
Acme thread	$H = P \cdot 0.5 + 0.010$ clearance
Stub acme	$H = P \cdot 0.3 + 0.10$ clearance
Square thread	$H = P / 2$
Worm thread	$H = 0.6866 \cdot P$
Buttress thread	$H = 0.66271 \cdot P$
Metric thread	$H = 0.541266 \cdot P$

Tapers

Finding taper per foot:

Taper-per-foot conversion to taper-per-inch: TPF / 12
Taper change on diameter: (TPF / 12) • L
Taper diameter on a small end: d + (TPF / 12) • L
Taper diameter on a large end: D – (TPF / 12) • L

TPF	taper per foot
L	length of taper
D	diameter on a large end
d	diameter on a small end

Finding taper in degrees:

Taper length: $\dfrac{D - d}{2 \cdot \tan \alpha / 2}$

Taper change on diameter: $\dfrac{D - d}{L}$

Taper diameter on a small end: $D - \dfrac{D - d}{L}$

Taper diameter on a large end: $d + \dfrac{D - d}{L}$

- α included angle
- L taper length
- D diameter on a large end
- d diameter on a small end

Finding metric taper:

Taper change on diameter: R • L
Diameter on a small end: D – (R • L)
Diameter on a large end: d – (R • L)

- R metric taper ratio (For example, 1:16 = 1/16 = 0.0625.)
- L length of taper
- D diameter on a large end in mm
- d diameter on a small end in mm

Converting taper ratio from inch data to metric data and vice versa:

$$\text{Metric ratio} = \frac{1}{\text{TPF} / 12}$$

$$\text{TPF} = \frac{12}{\text{metric ratio}}$$

Right triangle:

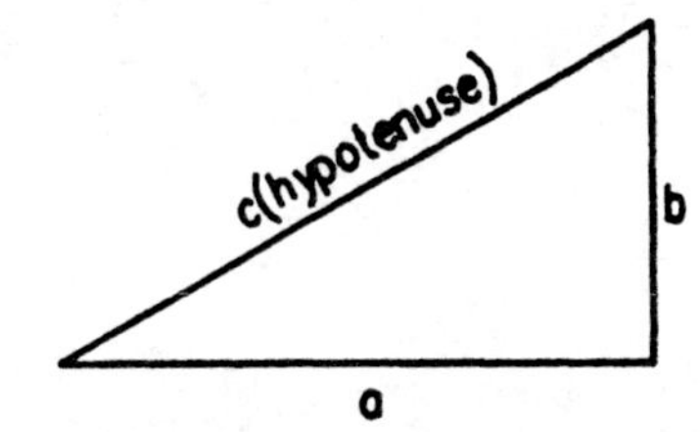

$$c = \sqrt{a^2 + b^2}$$

$$a = \sqrt{c^2 - b^2}$$

$$b = \sqrt{c^2 - a^2}$$

Trigonometric functions:

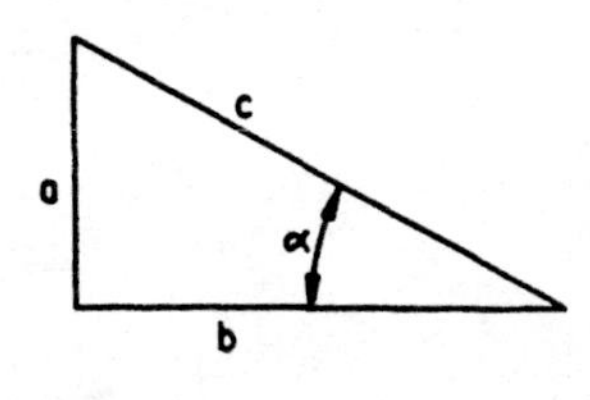

$$\text{SIN}\alpha = \frac{\text{opposite side}}{\text{hypotenuse}} = \frac{a}{c}$$

$$\text{TAN}\alpha = \frac{\text{opposite side}}{\text{adjacent side}} = \frac{a}{b}$$

$$\text{COS}\alpha = \frac{\text{adjacent side}}{\text{hypotenuse}} = \frac{b}{c}$$

$$\text{CTG}\alpha = \frac{\text{adjacent side}}{\text{opposite side}} = \frac{b}{a}$$

Appendix C
Figuring Compensation Amount When Chamfering

For Imaginary Tool Tip Programming

ANGLE	AXIS	TOOL NOSE RADIUS		
		1/64	1/32	3/64
15	X	0.0037	0.0073	0.011
	Z	0.0136	0.0271	0.0408
30	X	0.0066	0.0132	0.0199
	Z	0.0115	0.0229	0.0344
37.5	X	0.0079	0.0158	0.0238
	Z	0.0103	0.0206	0.031
45	X	0.0091	0.0183	0.0275
	Z	0.0091	0.0183	0.0275
52.5	X	0.0103	0.0206	0.031
	Z	0.0079	0.0158	0.0238
60	X	0.0115	0.0229	0.0344
	Z	0.0066	0.0132	0.0199

ANGLE	AXIS			
75	X	0.0136	0.0271	0.0408
	Z	0.0037	0.0073	0.011

For Tool Nose Center Programming

ANGLE	AXIS	TOOL NOSE RADIUS		
		1/64	**1/32**	**3/64**
15	X	0.012	0.024	0.036
	Z	0.002	0.0041	0.0062
30	X	0.009	0.018	0.027
	Z	0.0042	0.0084	0.0126
37.5	X	0.0077	0.0154	0.023
	Z	0.0053	0.0106	0.016
45	X	0.0065	0.013	0.0195
	Z	0.0065	0.013	0.0195
52.5	X	0.0053	0.0106	0.016
	Z	0.0077	0.0154	0.0232
60	X	0.0042	0.0084	0.0126
	Z	0.009	0.018	0.027
75	X	0.002	0.0041	0.0062
	Z	0.012	0.024	0.036

Appendix D
General Safety Notes for Safe CNC Operations

Safety is always a major concern in a metal-cutting operation. CNC equipment is automated and very fast, and consequently it is a source of hazards. The hazards have to be located and the personnel must be aware of them in order to prevent injuries and damage to the equipment. Main potential hazards include: rotating parts, such as the spindle, the tool in the spindle, chuck, part in the chuck, and the turret with the tools and rotating clamping devices; movable parts, such as the machining center table, lathe slides, tailstock center, and tool carousel; errors in the program such as improper use of the G00 code in conjunction with the wrong coordinate value, which can generate an unexpected rapid motion; an error in setting or changing the offset value, which can result in a collision of the tool with the part and/or the machine; and a hazardous action of the machine caused by unqualified changes in a proven program. To minimize or avoid hazards, try the following preventive actions:

1. Keep all of the original covers on the machine as supplied by the machine tool builder.
2. Wear safety glasses, gloves, and proper clothing and shoes.
3. Do not attempt to run the machine before you are familiar with its control.
4. Before running the program, make sure that the part is clamped properly.
5. When proving a program, follow these safety procedures:
    - Run the program using the machine Lock function to check the program for errors in syntax and geometry.
    - Slow down rapid motions using the RAPID OVERRIDE switch or dry run the program.

- Use a single-block execution to confirm each line in the program before executing it.
- When the tool is cutting, slow down the feed rate using the FEED OVERRIDE switch to prevent excessive cutting conditions.

6. Do not handle chips by hand and do not use chip hooks to break long curled chips; program different cutting conditions for better chip control. Stop the machine if you need to properly clean the chips.
7. If there is any doubt that the insert will break under the programmed cutting conditions, choose a thicker insert or reduce feed and/or depth of cut.
8. Keep tool overhang as short as possible, since it can be a source of vibration that can break the insert.
9. When supporting a large part by the center, make sure that the hole center is large enough to adequately support and hold the part.
10. Stop the machine when changing the tools, indexing inserts, or removing chips.
11. Replace dull or broken tools and/or inserts.
12. Write a list of offsets for active tools, and clear (set to zero) the offsets for tools removed from the machine.
13. Do not make changes in the program if your supervisor has prohibited your doing so.
14. If you have any safety-related concerns, notify your instructor or supervisor immediately.

Appendix E
Answers to Chapter Self-Tests

Chapter 1

1. Numerical Control (NC)
2. Computer Numerical Control (CNC)
3. Stepper motors
4. Direct current servomotors
5. Alternative current servomotors
6. An open-loop control system
7. A closed-loop control system
8. Automatic tool changers
9. random tool selection
10. sequential tool selection
11. Positioning
12. contouring
13. A rectangular coordinate system
14. a polar coordinate system
15. machine origin
16. part origin
17. program origin
18. EIA code set
19. ISO code set
20. Leading zero suppression
21. Trailing zero suppression

22. end of block code
23. Preparatory codes
24. Miscellaneous codes
25. Modal codes
26. Nonmodal codes

Chapter 2

1. absolute programming
2. incremental programming
3. Parameters
4. slash code
5. coordinate system preset
6. spindle speed limit
7. Tool offset
8. constant RPM
9. constant surface speed
10. Feed rate per spindle revolution
11. Feed rate per time
12. Linear interpolation
13. circular interpolation
14. arc center displacement
15. multiquadrant circular interpolation
16. dwell
17. subprogram
18. Nesting
19. direct numerical control (DNC)
20. machine Lock function
21. the Dry Run function

Chapter 3

1. home position
2. The Zero Return function
3. The master tool
4. work coordinates
5. presetting the register
6. manual data input (MDI)
7. tool length offsets
8. geometry offsets
9. tool reference point

10. imaginary tool tip programming
11. tool nose center programming
12. the measure function
13. A setup sheet
14. Work Zero Offset function
15. Relative position
16. Absolute position
17. Machine position
18. Distance to Go

Chapter 4

1. tool radius compensation
2. command point
3. hypotenuse
4. opposite side
5. adjacent side
6. sine, cosine, tangent, and cotangent

Chapter 5

1. automatic tool radius compensation
2. tool nose vector
3. compensated position
4. uncompensated position
5. Compensation left
6. Compensation right
7. Compensation cancel
8. a safety line

Chapter 6

1. single-point tools
2. Multipoint tools
3. High-speed steel (HSS)
4. Cemented carbide
5. Ceramics
6. Polycrystalline tooling
7. Cubic Boron Nitride (CBN)
8. round inserts

9. square inserts
10. triangular inserts
11. diamond-shaped inserts
12. hexagon-shaped inserts
13. Cutting fluid
14. operating conditions
15. The metal removal rate
16. cutting speed
17. The feed rate
18. Adaptive control

Chapter 7

1. Canned cycles
2. Fixed cycles
3. multiple repetitive cycles
4. acceleration
5. Deceleration
6. thread root diameter
7. spring pass
8. Right-hand threads
9. Left-hand threads
10. A single-lead thread
11. A double-lead thread
12. A multistart thread
13. Tool rake
14. zero and negative rake
15. Positive rake
16. peck grooving cycle

Chapter 8

1. drilling depth
2. tapping depth
3. Deep hole drilling
4. reaming
5. Countersinking
6. initial point level
7. R point level

8. Chatter
9. Internal vibrations
10. devibration boring bars
11. Milling
12. vibration
13. Helical interpolation
14. helix
15. fourth axis

Chapter 9

1. macro
2. automated
3. parametric programming
4. variable programming
5. macro program call
6. a macro body
7. argument
8. argument list
9. variable list
10. Control instructions
11. branching instructions
12. unconditional branch instruction
13. GOTO
14. conditional branch instruction
15. IF conditional branch
16. conditional expressions
17. program loop
18. looping
19. DO loop
20. WHILE conditional expressions
21. END statement

Chapter 10

1. Hardware
2. Software
3. MDI function
4. Memory function
5. Edit mode
6. RAM (Random Access Memory)

7. ROM (Read Only Memory)
8. SEARCH key
9. ALTER key
10. DELETE key
11. INSERT key
12. Secondary storage
13. application software
14. System software
15. floppy disk
16. hard disk
17. CNC program file
18. Part programming software
19. APT and COMPACT II
20. postprocessor

Glossary

A axis Machine axis that provides rotary motion of a part around the X axis; commonly assigned to the rotating head on the machining center.

absolute programming A programmed dimension given as the distance from the part origin. The sign for each coordinate depends on where the tool is moving according to quadrants.

accuracy The trueness of the measured value in relation to a known value.

adaptive control An automatic response of the machine control based on the parameter settings. It allows the control to adjust cutting conditions during machining.

address A letter of the alphabet used to define the meaning of the number that follows the address in a block of information.

APT (Automatically Programmed Tool) A computer programming language based on the concept that two reference surfaces, called the *part* and the *drive,* guide the tool toward a third reference surface called a *check surface*.

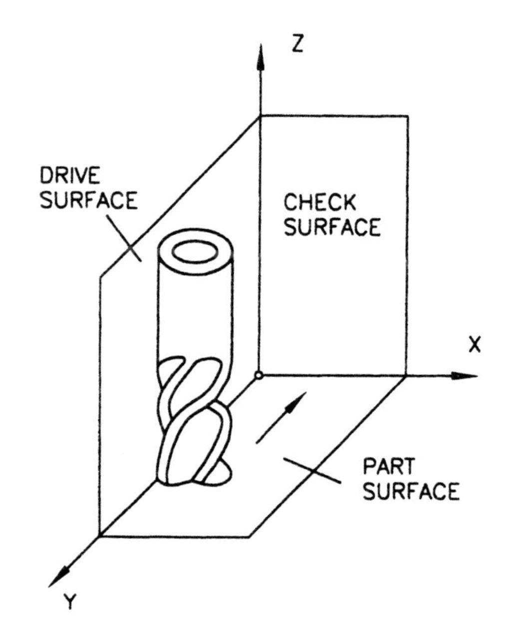

APT programming concept.

arc A circular motion of tool travel in either clockwise or counterclockwise direction in one of the plane pairs: X-Y, X-Z, or Y-Z.

arc center displacement The incremental distance from the arc start point to the arc center, represented by the addresses I, J, and K.

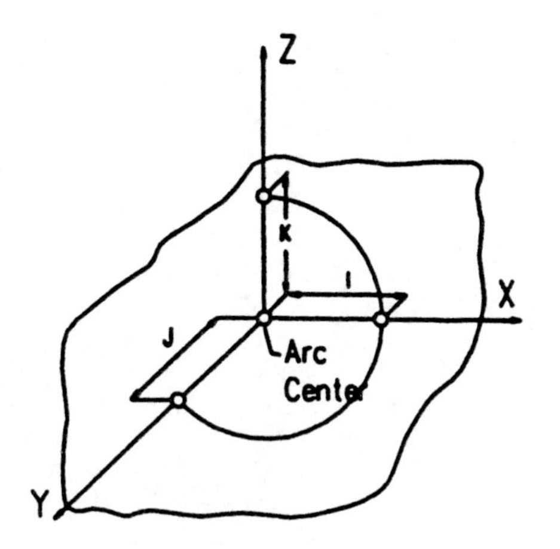

Arc center displacement.

arc clockwise (G02)/arc counterclockwise (G03) An arc generated by a simultaneous tool motion on two axes that is initialized by the preparatory codes G02 when cutting a clockwise arc or G03 when cutting a counterclockwise arc.

ASCII (American Standard Code for Information Interchange) This code is the most prevalent CNC programming system used today.

automatic tool changers Magazines on machining centers and turrets on the lathes that allow tool change without the intervention of the operator.

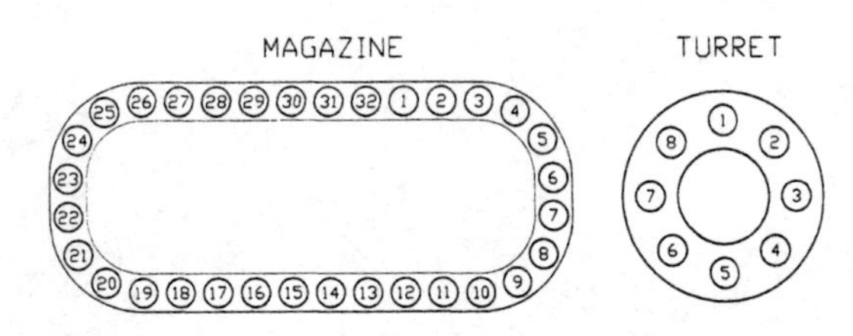

The automatic tool changers.

axis (1) One of the reference lines of a coordinate system; (2) A principal direction of the machine along which the movement of the tool or part occurs.

baud rate A signalling speed at which data is transmitted.

B axis Machine axis that provides rotary motion of a part around the Y axis; commonly assigned to the rotating head on the machining center.

binary A binary system that uses only the numbers 0 and 1 and a base 2 when expressing numbers.

BCD (Binary-Coded Decimal) A coding system that uses the binary system for representation of alphanumeric characters and numbers.

block A set of words, characters, or digits handled as a unit on one line of commands in a program. Blocks are separated by the end of block code character.

block delete A slash code (/) that permits selected blocks of a CNC program to be ignored (not executed) by the control. Whether or not a particular block will be deleted depends on the position of the BLOCK DELETE switch on the operator panel: if it is on, the information after the slash code will be ignored; if it is off, the information will be executed.

CAD (Computer-Aided Design) A process in which a designer uses a computer to create and/or modify a design, most often a drawing.

CAM (Computer-Assisted Manufacturing) A process in which a programmer uses a computer to define a manufacturing process, most often CNC programming.

cancel A command that cancels a canned cycle (G00 or G80) or tool radius compensation (G40).

canned cycle A preprogrammed sequence of events initiated by a single command. Its purpose is to reduce the amount of programming blocks needed to accomplish a particular task. This allows single-line programming of such common operations as threading, roughing, finish cutting, drilling, boring, and tapping.

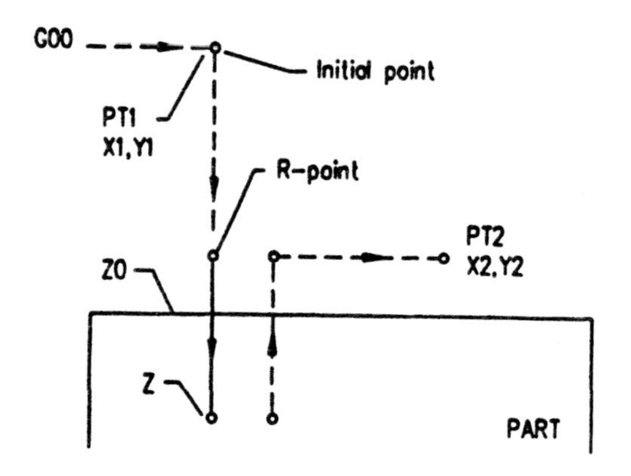

Tool motions in a canned cycle.

carbide A material composed of carbon and metal.

Cartesian coordinate system A rectangular coordinate system used to define distances from the origin in respect to the axes. This system, which is related to absolute programming, is created by perpendicular lines that intersect at the point of origin. The axes are X, Y, and Z.

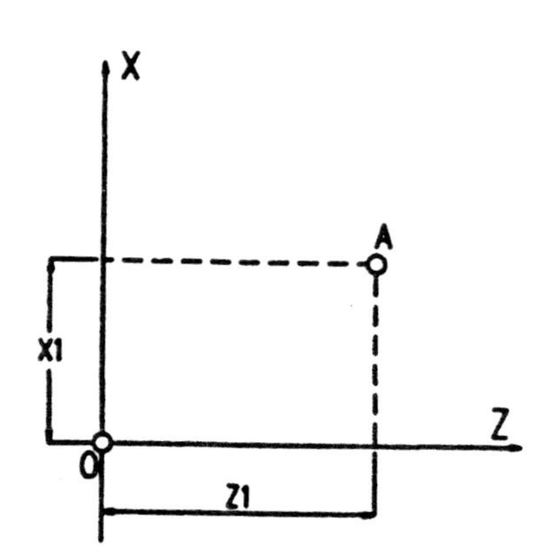

Cartesian coordinate system on the lathe.

CBN (Cubic Boron Nitride) A tooling material that can be used at speeds and feeds two to three times higher than those with carbide tooling. CBN tooling may last more than fifty times longer than carbide tools.

cemented carbides A form of carbide composed of tungsten carbide or titanium carbide and cobalt as a bond. It is used as cutting material in a variety of shapes of indexable inserts when cutting with higher cutting speeds and contact temperatures.

ceramics Cemented oxide tool materials made by sintering that are harder than carbides and allow cutting at higher temperatures and cutting speeds.

channel A communication path for data transmission.

character (1) One of a set of symbols; (2) A coded representation of symbols.

circular interpolation (G02, G03) A mode of contouring control that produces an arc of a circle or a full circle. In circular interpolation, the tool travels with simultaneous motion along two axes. The common codes for this mode are G02 for a clockwise arc, and G03 for a counterclockwise arc.

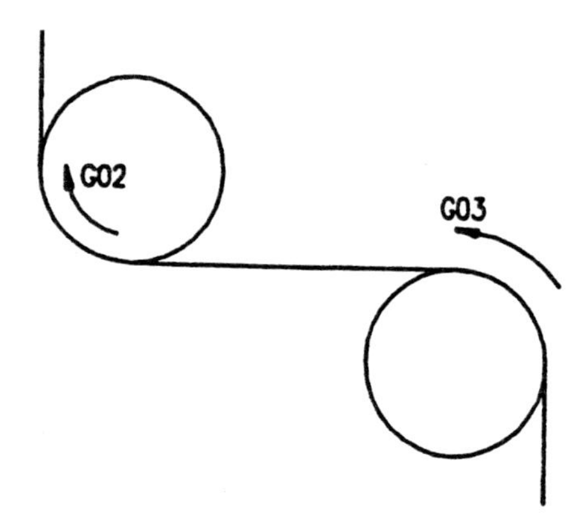

Circular interpolation.

command point A destination point of tool travel.

compensation cancel (G40) A command that cancels the compensation initialized by the G41 or G42 code. The control then returns to the uncompensated mode and the tool moves directly on the programmed shape.

compensation left (G41) A command that shifts the tool to the left of the part surface initialized by the G41 code.

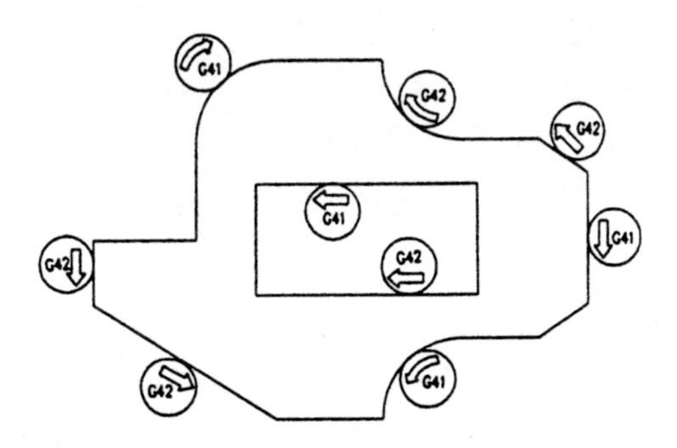

Compensation left/right.

compensation right (G42) A command that shifts the tool to the right of the part surface initialized by the G42 code.

computer A machine for processing data. In CNC programming, the computer is used with or without part programming software.

CNC (Computer Numerical Control) Numerical control performed under computer supervision.

constant surface speed A cutting speed generated by the speed control unit, which changes according to the part diameter being cut; the smaller the diameter, the more RPM is achieved; the bigger the diameter, the less RPM is commanded, which is why this mode is also known as the diameter speed.

continuous path An operation performed under continuous numerical control without an interruption in reading data.

contouring A cutting operation that results from simultaneous control of two or more axes. During this operation, the tool is in constant contact with the part. While contouring, the tool moves by the programmed feed rate.

cursor A visual movable pointer on the monitor which indicates the point of control. At this point you may change the content of computer memory.

cutter radius compensation A system that uses geometry tool offsets to affect the part size using the same tool diameter, to make it easier to set up tools of a different diameter from the original, and to achieve the roughing and finishing cuts using the same programmed data.

cutting speed The rotational speed of the cutting tool (on mills) or part (on lathes). It can be stated as constant revolution per minute (RPM) or constant surface feet per minute (SFM).

cycle A repeating sequence of operations.

cycle time The time required for one machining operation to be completed.

debug To check, locate, and correct mistakes in a computer program.

depth of cut The thickness of material removed in one pass of the tool.

dwell A programmed delay in program execution.

edit To modify a program or document.

EIA code set (Electronics Industries Association code set) An older coding system used for tape punching.

EOB (End of block code) A special character that separates the blocks of information in a CNC program. It is represented with the (;) character in the ISO code set, or the EOB in the EIA code set.

end of program (M02) A miscellaneous function that stops the spindle, coolant, and feeding but cannot rewind the tape to the beginning.

F address An address used to specify a feed rate amount.

fixed canned cycles These cycles allow programming three or four successive tool movements in a single block.

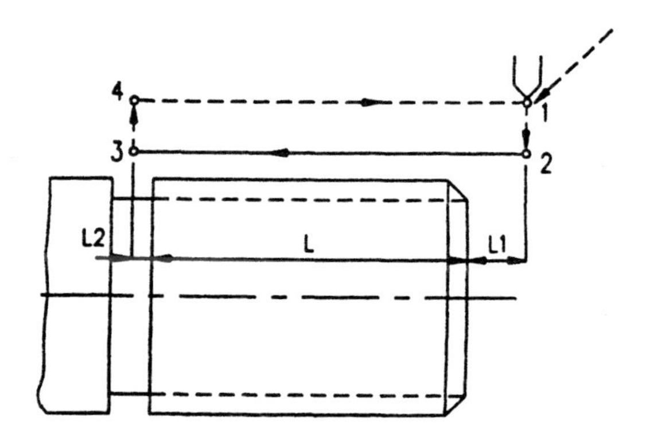

Fixed canned cycle.

fixed sequential block format An older coding format that required data to be programmed in a particular sequence.

fixed zero machines Older machines which have a permanent zero location that does not allow direct change in location of the part origin.

floating zero The ability of a control unit to establish a part origin at any point inside the electronic and mechanical limits of the machine. Almost all newer machines are designed to allow full zero shifting.

format A specific set of rules for entering, receiving, and outputting data, also known as CNC syntax, programming format, or tape format.

G codes Preparatory codes designated by the address G used to prepare the machine to treat programming information in a distinct manner and to execute it. In essence, the G codes decide the mode of the system.

geometry offset An offset used to compensate for tools that differ in diameter.

hard copy A printout of a CNC program.

hardware Any physical component in a computer system, such as the processor, monitor, memory, and input/output devices.

H code A code used to assign the length offset number on the machining center.

helical interpolation This feature allows circular interpolation on two axes simultaneously, while providing linear cutting on a third axis.

home position The point from which the tools start program execution and to which they return after completion of the machining sequence.

HSS (high-speed steel) An alloy steel that contains primarily tungsten and cromium with a small percentage of cobalt, vanadium, and molybdenum. It is used as cutting material mainly for end mills and drills when cutting with lower speeds and contact temperatures.

imaginary tool tip An imaginary point on the tool at the intersection of the lines tangent to the tool nose radius.

imaginary tool tip programming A type of programming that uses the imaginary tool tip as a reference point.

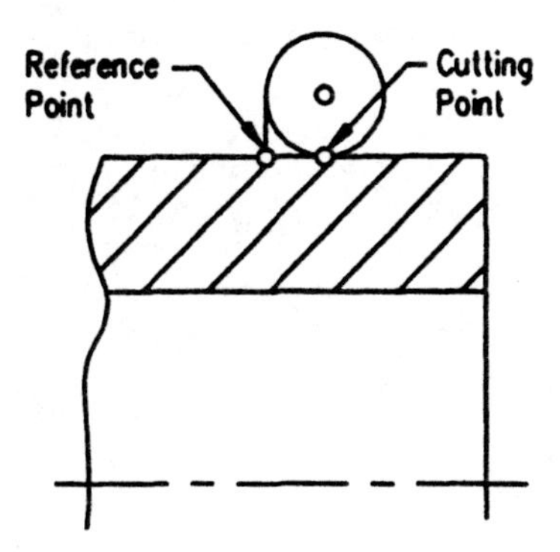

Imaginary tool tip programming.

incremental programming Programmed movement based on the change in position between two successive points.

interpolation The ability of a control to generate a simultaneous movement of two or more axes in order to produce a defined geometric pattern, such as a linear, circular, or helical pattern.

IPM (inches per minute) Used to designate a feed rate per time, usually on mills.

IPR (inches per revolution) Used to designate a feed rate per spindle revolution, usually on lathes.

leading edge programming (*See* imaginary tool tip programming.)

leading zero suppression A programming format that eliminates the need for programming zeros to the left of the first significant digit in a coordinate word.

linear interpolation The ability of a control to generate a simultaneous movement of two or more axes in order to produce a linear (straight-line) path.

looping The ability of a control to repeat a specified instruction a specified number of times.

machine origin The origin point of the machine coordinate system, which is set by the machine tool builder and normally cannot be changed.

machining center The CNC machine tool capable of performing multiple operations in one setup. The machining center is normally equipped with an automatic tool changer.

macro Short for *macroinstruction,* a macro is a series of instructions that can be executed repeatedly. While repetition is in progress, a macro program repeats one or more programmed values. These changeable values are known as *parameters* or *variables*. Consequently, macro programming is also known as *parametric programming* or *variable programming*.

manual programming A means of programming a CNC machine using manual calculations and designing the program without part programming software.

M codes Codes used to perform miscellaneous functions, such as turning coolant on and off.

MDI (manual data input) A means of inserting data manually into the CNC control using a machine keyboard and switches.

modal A command retained in a system until canceled or replaced by a new command.

modal code A code that remains in effect until canceled by another code within the same group.

multiple repetitive canned cycles These cycles allow repeating any number of passes to cut material in repeating sequences until the specified profile is achieved.

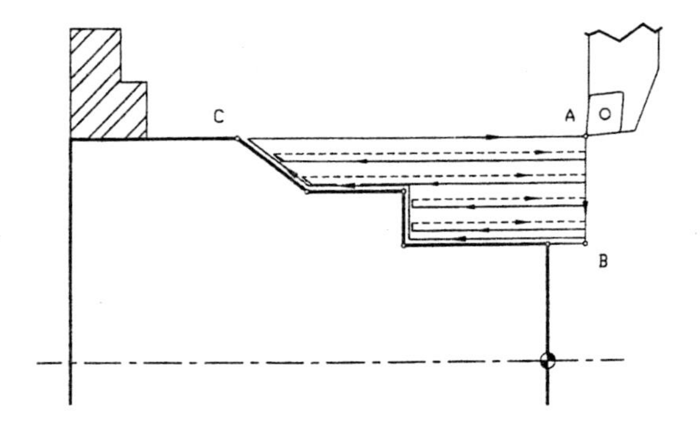

Multiple repetitive canned cycle.

multiquadrant circular interpolation Cutting an arc in more than one quadrant.

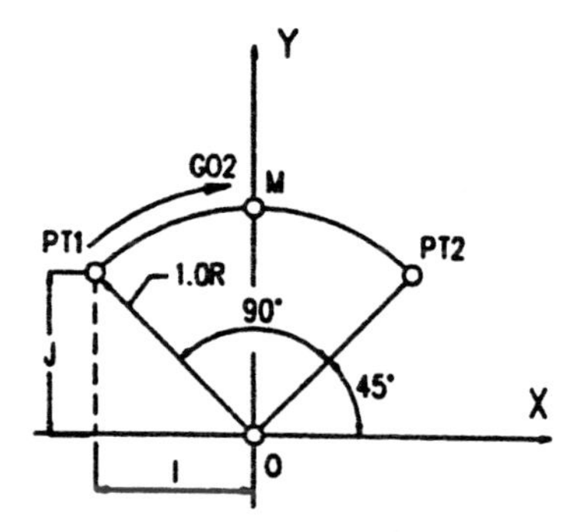

Multiquadrant circular interpolation.

NC (numerical control) Any machining process in which the operations are performed automatically in sequences as specified by a program that contains information for tool movements.

nonmodal code A code that remains in effect for the block programmed.

offset A compensation in slide movement necessary to accommodate dimensional variations of tools.

optional stop (M01) Similar to the program stop command, this miscellaneous command is used to stop the program execution at the operator's discretion. Whether or not the program will stop depends on the position of the OPTIONAL STOP switch. If the switch is on, the program will stop. If the switch is off, execution continues without interruption. To continue the program execution after stopping, the operator must press the CYCLE START button.

origin A zero reference point of the coordinate system. There are three origins: part origin, machine origin, and program origin.

parameter (1) A machine parameter used to set the system; (2) A variable in a macro program; (3) A constant in a canned cycle.

part origin The origin from which absolute coordinates are measured.

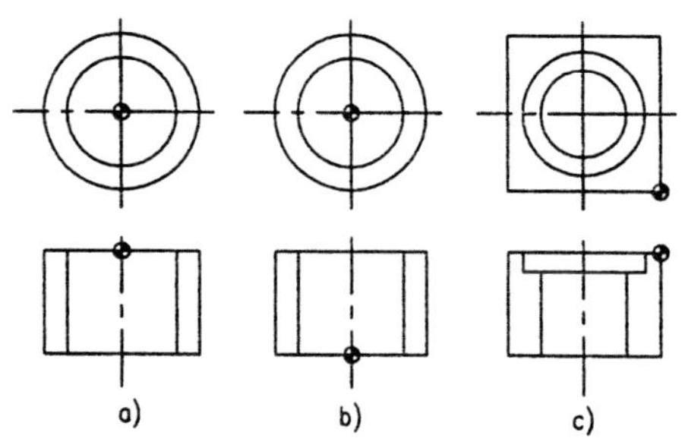

The part origin.

PCD (polycrystalline) diamond A synthetic diamond used for tooling, mainly used for finish operations on nonmetallic materials.

plane A two-dimensional surface in a rectangular coordinate system.

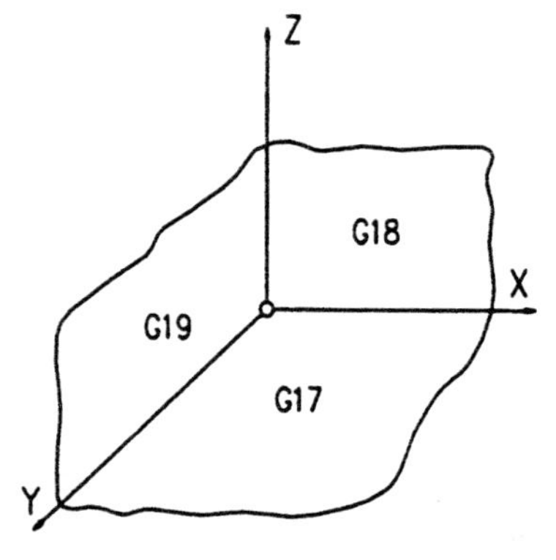

The planes of the rectangular coordinate system.

point-to-point programming Positioning of the tool from one point to another when programming such operations as drilling, tapping, boring, and reaming. During this process, the tool is not in constant contact with the part, and when changing locations, the tool is moved by rapid traverse rate.

polar coordinate system A coordinate system used on some machining centers in which a coordinate point is described by the radius vector and angle.

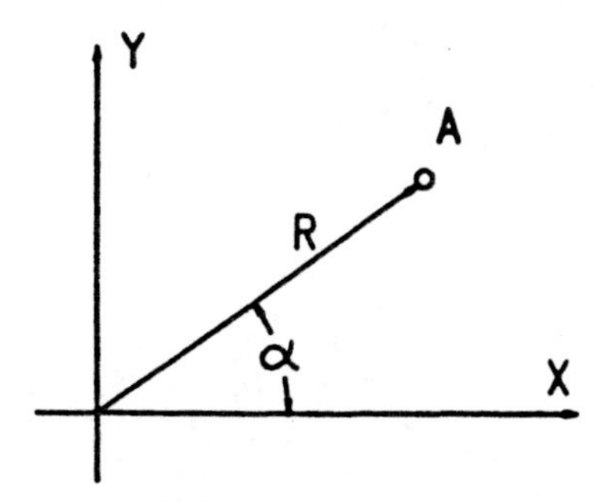

The polar coordinate system.

program A set of coded instructions to complete a particular job.

programmed dwell (G04) Delay in program execution for a programmed length of time.

program origin (G50, G92) The tool start point in the program.

program stop (M00) A miscellaneous function used to stop program execution. When it comes into effect, the spindle, coolant, and feeding stop. To continue, the operator has to press the CYCLE START button.

quadrant One of the four sections of the rectangular coordinate system.

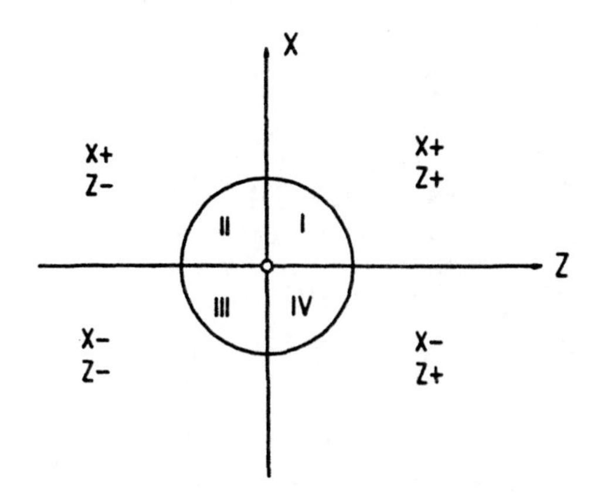

Quadrants of the rectangular coordinate system.

RAM (random access memory) Volatile memory, the content of which can be altered. Used to store programs, when power is turned off, the content of this memory is lost if batteries are not used. Also known as Read/Write memory.

rapid traverse (G00) A programmed movement at the maximum rate of machine travel. The rate is set by a binary number.

rectangular coordinate system (*See* Cartesian coordinate system.)

register A memory location used for temporary storage of data, such as a tool offset value.

resolution The smallest amount of movement along an axis that a CNC machine can perform.

ROM (read only memory) Nonvolatile memory, from which one can read data but to which one cannot write. Machine software is stored in this memory for permanent use. The contents of this memory do not change when the power is turned off.

RPM Spindle speed expressed in revolutions per minute.

sequence number Block identification number (N) in a CNC program. In some instances it must be specified in order to enable the control to execute the program properly, such as when using the G71 canned cycle or calling a subprogram.

setup Preparation of a machine for the machining process by means of loading tools and fixtures, establishing the origin points, and producing the first quality part according to specifications.

SFM (surface feet per minute) (*See* constant surface speed.)

significant digit A digit that must be kept to preserve the accuracy of the quantity.

software A general term for programs stored in electronic form.

subroutine A separate program related to the main program that allows repetition of the programmed routine. A subroutine can be a subprogram or a macro program.

tab sequential format An older tape format that used tab codes to specify the word address.

tool length offset An offset used to compensate for tools that differ in length. It allows programming imaginary tool lengths.

tool nose center programming A type of programming that uses the tool nose center as the reference point.

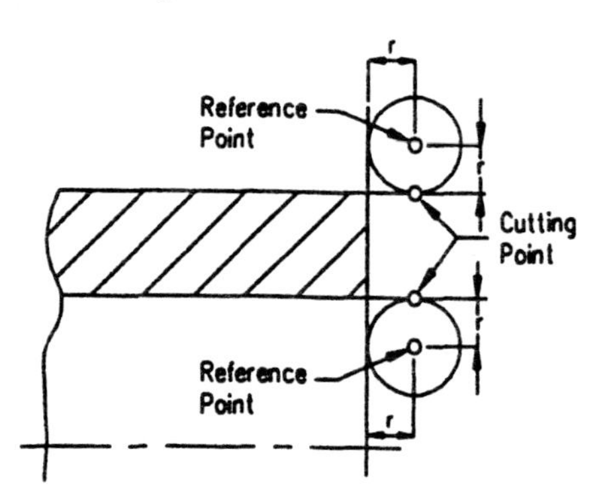

Tool nose center programming.

tool reference point The point on the tool used to position the tool when programming.

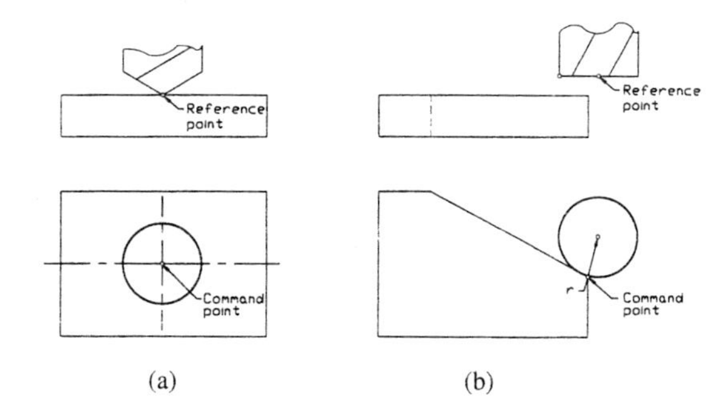

Tool reference point.

touch-off A method of setting up the tool by means of touching-off the part surface by the tool tip.

trailing zero suppression A programming format that eliminates the need for programming zeros to the right of the first significant digit in a coordinate word.

vector A quantity that has magnitude and direction, used to describe the coordinates of a point.

word A combination of a letter and numerals, such as Z1.500.

X axis The line parallel to the machine table on the machining center, and perpendicular to the spindle centerline on the lathe.

Y axis The line perpendicular to the X axis on the machining center.

Z axis The line parallel to the spindle centerline on either the lathe or the machining center.

Index